Springer Theses

Recognizing Outstanding Ph.D. Research

Aims and Scope

The series "Springer Theses" brings together a selection of the very best Ph.D. theses from around the world and across the physical sciences. Nominated and endorsed by two recognized specialists, each published volume has been selected for its scientific excellence and the high impact of its contents for the pertinent field of research. For greater accessibility to non-specialists, the published versions include an extended introduction, as well as a foreword by the student's supervisor explaining the special relevance of the work for the field. As a whole, the series will provide a valuable resource both for newcomers to the research fields described, and for other scientists seeking detailed background information on special questions. Finally, it provides an accredited documentation of the valuable contributions made by today's younger generation of scientists.

Theses are accepted into the series by invited nomination only and must fulfill all of the following criteria

- They must be written in good English.
- The topic should fall within the confines of Chemistry, Physics, Earth Sciences, Engineering and related interdisciplinary fields such as Materials, Nanoscience, Chemical Engineering, Complex Systems and Biophysics.
- The work reported in the thesis must represent a significant scientific advance.
- If the thesis includes previously published material, permission to reproduce this must be gained from the respective copyright holder.
- They must have been examined and passed during the 12 months prior to nomination.
- Each thesis should include a foreword by the supervisor outlining the significance of its content.
- The theses should have a clearly defined structure including an introduction accessible to scientists not expert in that particular field.

More information about this series at http://www.springer.com/series/8790

Robert J. Lewis-Swan

Ultracold Atoms for Foundational Tests of Quantum Mechanics

Doctoral Thesis accepted by
The University of Queensland, Brisbane, Australia

 Springer

Author
Dr. Robert J. Lewis-Swan
School of Mathematics and Physics
The University of Queensland
Brisbane
Australia

Supervisor
A/Prof. Karén Kheruntsyan
School of Mathematics and Physics
The University of Queensland
Brisbane
Australia

ISSN 2190-5053   ISSN 2190-5061   (electronic)
Springer Theses
ISBN 978-3-319-82252-5   ISBN 978-3-319-41048-7   (eBook)
DOI 10.1007/978-3-319-41048-7

Printed on acid-free paper

This Springer imprint is published by Springer Nature
The registered company is Springer International Publishing AG Switzerland

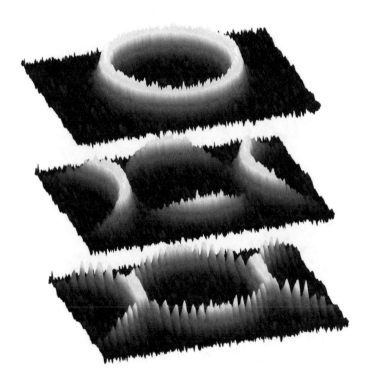

Parts of this thesis have been published in the following journal articles:

1. R.J. Lewis-Swan and K.V. Kheruntsyan *Sensitivity to thermal noise of atomic Einstein-Podolsky-Rosen entanglement.* Phys. Rev. A **87**, 063635 (2013)

2. R.J. Lewis-Swan and K.V. Kheruntsyan *Proposal for demonstrating the Hong-Ou-Mandel effect with matter waves.* Nature Communications **5**, 3752 (2014)

3. R.J. Lewis-Swan and K.V. Kheruntsyan *Proposal for a motional-state Bell inequality test with ultracold atoms.* Phys. Rev. A **91**, 052114 (2015)

Supervisor's Foreword

In his thesis, Robert Lewis-Swan has investigated a variety of topics related to the generation, characterization, and exploitation of quantum correlations and entanglement among ultracold atoms. Specifically, it is focused on these quantum phenomena in two contexts: (i) demonstrating nonclassical correlations, e.g., violations of a Bell inequality and the Einstein–Podolsky–Rosen (EPR) paradox, with ensembles of ultracold atoms, and (ii) improvements in quantum technologies by exploitation of entanglement, such as quantum-enhanced metrology. In particular, the work presented in this thesis emphasizes the possibility to demonstrate and characterize entanglement in realistic experiments, beyond simple "toy-models" often discussed in the literature.

The thesis can be broken into three main sections. The first contains a pair of important theoretical proposals to experimentally realize an atomic Hong–Ou–Mandel (HOM) effect and a violation of a motional-state Bell inequality for the first time. Each proposal is built around the process of atom-pair production via elastic s-wave scattering from a pair of colliding Bose–Einstein condensates, which has been characterized extensively in recent experimental research. Robert utilized the quantum correlations between pairs of atoms produced via this process, in combination with appropriate operational measures, to define and characterize the HOM effect and a Bell inequality in this multimode atomic system. While his modeling extensively incorporates the details of a realistic experimental setup, he was able to connect the complex results to simplified models based only on knowledge of the density–density correlations of the system, allowing important insights for laboratory realizations of these quantum phenomena. The relevance of this work was later reflected in the demonstration of the atomic HOM effect in the lab of Alain Aspect and Chris Westbrook at the Institut d'Optique.

The second section of the thesis follows the theme of characterizing and understanding realistic limitations of experimental schemes, by examining the effects of thermal fluctuations in spin-changing collisions in spinor condensates, and hence clarifying the role of these fluctuations in the possible demonstration of the EPR paradox with massive particles. This work was in direct response to an earlier reported experimental measurement of the EPR criterion by Markus Oberthaler's experimental group at Heidelberg University. These experiments suffered from large technical noise, which prevented a definitive observation of the EPR paradox; Robert's research identified at least one source of such noise—thermal fluctuations—while also characterizing the relative robustness of various other measures of entanglement and nonclassicality to thermal fluctuations.

The final section is focused on how quantum correlations can be exploited for quantum technologies, specifically quantum-enhanced metrology in the first experimental realization of an atomic SU(1,1) interferometer. This work was a collaborative effort with the Oberthaler group, driven by Robert's prior research on spin-mixing in spinor condensates. Generically, the SU(1,1) quantum-enhanced interferometer exploits the nonclassical correlations produced by a pair-production process, combined with an unconventional nonlinear readout scheme using time-reversal of the same process. The theoretical work presented in this thesis particularly focuses on the implementation of such an interferometer utilizing spin-mixing in a spinor condensate as the pair-production process, and the associated subtleties of such a realization.

The attraction of Robert's research on these broad topics is the relative simplicity of the models and analysis while having direct applications to experimentally feasible setups. This is particularly impressive given that the experiments under consideration typically involve interacting quantum many-body systems composed of a large number of atoms. The quality, originality, and relevance of this work is reflected in a spate of recent publications regarding experimental demonstrations of, e.g., the atomic Hong–Ou–Mandel effect, observation of EPR entanglement with massive particles in the Hannover group of Carsten Klempt, and the demonstration of an atomic SU(1,1) interferometer. This demonstrates that the work is at the forefront of current theoretical research in ultracold atomic physics. As such, it is only appropriate that Robert's outstanding research is recognized by publication in the renowned Springer Theses series and I hope it will serve as a useful resource for those interested in the field of quantum atom optics.

Brisbane, Australia   A/Prof. Karén Kheruntsyan
April 2016

Abstract

We investigate the generation, characterization and measurement of nonclassical correlations and entanglement in ultracold atomic gases. Specifically, we propose new tests to demonstrate nonclassical correlations, Einstein–Podolsky–Rosen (EPR) entanglement and Bell inequality violations, in systems involving dilute gas Bose–Einstein condensates (BECs). We focus on the challenges of generating and preserving these correlations in atom optics schemes with massive particles and define appropriate operational measurements to demonstrate them experimentally. In doing so, we characterize how measures of EPR entanglement and violations of a Bell inequality evolve with time and scale with system size. Further, we also investigate practical applications of nonclassical correlations and entanglement to quantum-enhanced technology, specifically quantum metrological schemes.

Acknowledgments

First, I wish to thank my supervisor Karén Kheruntsyan for his fantastic support over the past four and a half years. I am very thankful for the opportunities he has provided me throughout this period and the time and effort he has put into my education as a physicist. His approach to physics has taught me a great deal and I can only hope to aspire to his example. I should also mention that his persistent approach to proofreading played no small role in making this thesis and the papers herein the work that it is.

I am also thankful to Joel Corney, Matt Davis, Simon Haine, and Stuart Szigeti for helping me with my countless questions. They provided invaluable assistance whenever I had a question that stumped me.

Special mention should also go to the experimental group of Heidelberg University, who welcomed me during my two visits. I thank them for the illuminating experience, which gave me a fantastic insight into the realities of experimental physics. Also, for the trustworthy guidance on good German beer.

To the other students who shared in this common pain: Jan Zill, James Mills, Nick McKay-Parry, Behnam Tonekaboni, and others. I thank everyone for the countless distractions, nerf darts to the face and beer which have kept me sane through this process, particularly in the times when I discovered *new physics*.

Lastly, I wish to thank my family and my fiancée Jen for their support and putting up with the quirks of a physicist. Without their understanding this thesis could not have come to fruition.

Above all, I would like to acknowledge the support of my late grandmother, Dame Mavis E. Lewis (OBE), who made all of this possible. Although you were not able to see me pass the final hurdle, I promise I will wear the silly hat and smile for the photographs.

Contents

Chapter 1
Introduction and Background Physics

Quantum mechanics has time and again proven to be the most accurate scientific model of the microscopic realm, specifically as a tool to predict the outcomes of experimental measurements. However, since the seminal 1935 paper of Einstein, Podolsky and Rosen (EPR) [1] fierce debate has raged between physicists over the philosophical consequences of quantum mechanics.

In particular, there has been argument as to whether quantum mechanics can be regarded as a complete physical theory of the microscopic reality, wherein one is forced to accept phenomena which are in stark contrast with our everyday experience of the macroscopic world, such as entanglement, non-locality (referred to by Einstein as "spooky-action-at-a-distance") and the Heisenberg uncertainty principle. Or, as EPR would have it, should we interpret these features as a smoking gun of an incomplete theory of nature, exhibiting our ignorance of the underlying 'true' microscopic reality? In response, one might seek to uncover so-called 'hidden variables' to supplement the current formalism and restore a classical notion of the universe, wherein, for instance, particles may have well-defined position and momentum simultaneously. Physicists are divided on this issue, although many are happy to take a third option and 'shut up and calculate' [2], simply regarding quantum mechanics as a useful toolbox with which to predict measurement outcomes.

Much of the philosophical unease surrounding this issue has its roots in the stark differences between the microscopic realm of quantum mechanics and the macroscopic classical world. In particular, an ongoing research question is how the classical world emerges from the microscopic rules of quantum mechanics. Specifically, why prevalent quantum effects such as entanglement and non-locality are not realized in our macroscopic realm. Some answers to these big questions can be found by investigating systems at the microscopic scale, and understanding why entanglement and non-local correlations are difficult to generate and preserve even at this scale.

To address these issues one should first begin by gaining an understanding of the two touchstones of the philosophical and physical debate: the Einstein–Podolsky–Rosen paradox and Bell inequalities.

© Springer International Publishing Switzerland 2016
R.J. Lewis-Swan, *Ultracold Atoms for Foundational Tests of Quantum Mechanics*, Springer Theses, DOI 10.1007/978-3-319-41048-7_1

1.1   The Einstein–Podolsky–Rosen Paradox

In 1935, Einstein, Podolsky and Rosen published their influential paper [1], in which they argued that quantum mechanics as a theory was incomplete. Their conclusion came as a consequence of their motivation to preserve what they termed local realism in any theory of the quantum world. They argued that any sensible theory must uphold locality, in the sense that for two space-like separated systems any measurement or action on one may not affect the reality of the other. Further, EPR argued that "if, without in any way disturbing a system, we can predict with certainty the value of a physical quantity, then there exists an *element of physical reality* (sic) corresponding to this physical quantity".

To illustrate their argument they devised a paradoxical situation in which they consider two particles, which we will label 1 and 2 respectively, entangled through some interaction at the origin such that they have perfectly correlated positions $x_1 - x_2 = u$ wher u is the center-of-mass position and sum momenta, $p_1 + p_2 = 0$. Due to these correlations, measuring the position of particle 1 enables one to infer the exact position of particle 2, even if the particles are space-like separated and therefore a local measurement on one particle may not influence the state of the other. Similarly a measurement of the momentum of particle 1 allows one to infer exactly the momentum of particle 2. Following the logic of EPR, the inferred values x_2 and p_2 must be regarded as pre-existing (elements of reality) for the quantum mechanical wavefunction to be regarded as a complete description of reality. However, if this is true then it holds that the product of uncertainties of the inferred quantities vanishes,

$$\Delta^2 x_2 \Delta^2 p_2 = 0 \qquad\qquad (1.1)$$

for the perfectly correlated state under consideration, where $\Delta^2 x_i = \langle \hat{x}_i^2 \rangle - \langle \hat{x}_i \rangle^2$ is the variance of the measurement of $\hat{x}_i$ and similar for $\hat{p}_i$. This appears to violate the Heisenberg uncertainty principle for a pair of canonically conjugate observables, which requires that

$$\Delta^2 x_2 \Delta^2 p_2 \geq \hbar^2/4 \qquad\qquad (1.2)$$

for $[\hat{x}_j, \hat{p}_j] = i\hbar$. In the standard interpretation of quantum mechanics, the response to this paradox is that it is impossible to perform simultaneous measurements of position and momentum and thus simultaneous elements of reality for these quantities may not exist. However this implies that the choice of measuring position or momentum affects the reality of the second particle and, as a consequence, one must accept that the theory of quantum mechanics is intrinsically non-local. Due to their motivation to preserve local realism, EPR concluded instead that quantum mechanics must be an incomplete description of reality which must be supplemented by extra 'hidden' variables.

In the spirit of preserving local realism, there have been many attempts to construct local hidden variable theories of quantum mechanics. Such theories attempt to prescribe extra variables to describe the missing elements of reality in the quantum mechanical formalism, and thus construct what EPR would define as a complete physical theory. A prominent example of such a theory was developed by Bohm [3], which reproduced the results of quantum mechanics for discrete spin variables whilst all the (canonically conjugate) spin components were able to be defined simultaneously. Intriguingly, however, it was later shown that this theoretical construct preserved the 'flaw' of non-locality. This issue was famously investigated by John Bell and motivated his work on Bell inequalities, which we shall discuss further in Sect. 1.2.

A notable experimental demonstration of the EPR paradox, in the sense of similarity to the original EPR construction with continuous variables $\hat{x}$ and $\hat{p}$, was performed by Ou et al. [4]. Their demonstration followed the theoretical construct proposed by Reid [5], and demonstrated a continuous-variable EPR paradox using optical quadratures of the signal/idler beams produced by optical parametric downconversion. Reid's formulation is an extended version of the EPR paradox for a system with imperfect correlations, in contrast to the perfectly correlated, but unphysical [6], EPR state. By measuring non-commuting quadratures of the signal beam, $\hat{X}_1$ and $\hat{Y}_1$, which are analogous to position and momentum operators, one is able to infer the value of the respective quadratures of the correlated idler beam, $\hat{X}_2$ and $\hat{Y}_2$, within some uncertainty $\Delta_{\text{inf}}^2 X_2$ and $\Delta_{\text{inf}}^2 Y_2$ respectively (the form of $\Delta_{\text{inf}}^2 X_2$ depends on the correlation between the modes, $\langle \hat{X}_1 \hat{X}_2 \rangle$, and similar for $\Delta_{\text{inf}}^2 Y_2$). A demonstration of the EPR paradox in this formulation then corresponds to a product of inferred quadratures, $\Delta_{\text{inf}}^2 X_2 \Delta_{\text{inf}}^2 Y_2 < 1$, whereas the Heisenberg uncertainty limit for the actual quadratures requires $\Delta^2 X_2 \Delta^2 Y_2 \geq 1$ as $[\hat{X}_i, \hat{Y}_i] = 1$.

Recently, attempts have been made to demonstrate the paradox with massive particles for the first time, in particular the experiment of Gross et al. [7] which used spin-changing collisions in a spinor Bose–Einstein condensate (BEC) to produce a state analogous to that of Ref. [4]. Measurements of the atomic field quadratures led to an inconclusive result, nevertheless, it was shown that the produced state was entangled, in the sense of inseparability of the state [8].

1.2 Bell Inequalities and Local Hidden-Variable Theories

In 1964, after noticing the issue of non-locality present in Bohm's hidden variable theory for discrete spin variables, John Bell derived a set of conditions which any prospective local hidden variable theory must obey [9]. Known as Bell inequalities, they could be directly compared to the predictions of quantum mechanics.

Consider two space-like separated measurements on correlated particles with outcomes A and B (with $A, B = \pm 1$ for simplicity), which only depend on the measurement settings a and b respectively, which could for instance be polarization settings of a photo-detector. Furthermore, we consider the theory to have an arbitrary set of hidden variables, λ, whose only constraint is that they are characterized by some probability distribution $\rho(\lambda)$. We may then write the joint probability of the measurement outcomes as

$$P(A, B|a, b) = \int d\lambda \rho(\lambda) P(A|a, \lambda) P(B|b, \lambda), \qquad (1.3)$$

where $P(Z|z, \lambda)$ is the conditional probability of a measurement result Z given the known setting z and hidden variable λ. Locality, in terms of the EPR argument, is present in the right-hand side of this expression by our explicit choice that the outcomes A and B can only depend on their respective local settings a and b. It may be shown that for a sum of these conditional probabilities

$$E(a, b) = P(1, 1|a, b) + P(-1, -1|a, b) - P(-1, 1|a, b) - P(1, -1|a, b), \qquad (1.4)$$

there exists the inequality,

$$S = |E(a, b) + E(a, b') + E(a', b) - E(a', b')| \le 2, \qquad (1.5)$$

which is known as the Clauser–Horne–Shimony–Holt (CHSH) version of a Bell inequality.

Bell was able to demonstrate that there exist certain entangled states in quantum mechanics which violate this inequality, by up to a factor of $\sqrt{2}$ (corresponding to $S = 2\sqrt{2}$). The existence of such entangled states thus implies that quantum mechanics has non-local correlations which cannot be explained by any local hidden variable theory. As a result, the central motivation and philosophical foundation of EPR's argument, local realism, is thus explicitly incompatible with the quantum world.

Importantly, Bell's result can be cast in terms of experimentally measurable quantities which can be used to rule in or out a description by a local hidden variable theory. Notable experiments demonstrating a violation of a Bell inequality were conducted by Aspect (1982) [10], Zeilinger (1998) [11], Rarity and Tapster (1990) [12], Wineland (2001) [13] and Sakai (2006) [14]. All of these experiments used pairs of entangled photons, except for the experiments of Wineland and Sakai, which used trapped ions and proton-proton pairs respectively. However, none of these experiments have overcome both of the detection and locality loopholes. In the case of the first loophole a minimum detector efficiency is required to differentiate between quantum mechanics and a local hidden variable theory (in the case of the above CHSH inequality this is $\sim 83\%$ [15]). The experiments of Wineland and Sakai overcame this, however they did not also overcome the second loophole, which requires the measurements to be space-like separated. This loophole was overcome by Zeilinger's experiment in 1998, in which the measurement apparatus of the respective photons was separated by a distance of 400 m and the detector settings were chosen during the time-of-flight of the photons, although the detector efficiencies were below the required 83 % (Fig. 1.1).

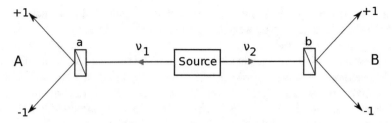

Fig. 1.1 Example of a Bell inequality test experiment. An entangled two-particle state is produced by a source, in this case a pair of photons with entangled polarization (labeled ν_1 and ν_2). We choose to perform measurements of type a and b, i.e. polarization angles of the detectors, with outcomes $A = \pm 1$ and $B = \pm 1$ respectively. A Bell inequality places certain constraints on the joint-detection probabilities of the measurement outcomes for a local hidden variable theory, which quantum mechanics is shown to violate

1.3 Motivation and Scope of This Thesis

Foundational tests of physics involving the EPR paradox and Bell inequalities have been focused primarily within the field of quantum optics. In part, this has been driven by the relative ease with which massless photons could be prepared in quantum states useful for such tests, such as the two-mode squeezed vacuum state produced by optical spontaneous parametric down-conversion. The motivation of this thesis is to push these foundational tests of quantum mechanics into new regimes, in particular, to generate and characterize non-classical correlations and entanglement between ensembles of massive particles. Such tests will allow us to gain insight into important research questions, such as:

- **What are the appropriate mechanisms and systems to generate EPR entanglement and non-local correlations in ultracold atomic gases?** In particular, how can we leverage prior knowledge gained from quantum optics to assist in identifying processes which will produce useful states, such as the two-mode squeezed vacuum state.
- **Why are large-scale entanglement and non-local correlations between massive particles difficult to generate and observe?** Importantly, we wish to better understand and separate technical and fundamental limitations in the context of experimental demonstrations of EPR entanglement and Bell inequalities.
- **What are appropriate operational measures to quantify EPR entanglement, non-classical correlations and violation of a Bell inequality in ultracold atomic gases?** Particularly, by characterizing the new and unique features of atom-optics schemes (in comparison to quantum optics) how can we appropriately quantify the degree of entanglement or non-classicality?
- **How does EPR entanglement and violation of a Bell inequality evolve with time and scale in system size?** Specifically, what can this insight tell us about the robustness of results derived from idealized analytic models when generalized to realistic experimental scenarios.

Answers to these questions can provide valuable insight and a better understanding of the ill-defined transition between the quantum and classical realms [16], particularly how mesoscopic and macroscopic physics may emerge from the microscopic constituents. Furthermore, pushing these fundamental tests into the realm of macroscopic ensembles of massive particles leads to coupling to gravitational fields, which in turn may provide insights into as yet un-established theories of quantum gravity [17]. Particularly, design of feasible experimental schemes may provide a first step to tests of modifications to the currently established theory of quantum mechanics, which is incompatible with gravitational theories based on General Relativity.

A better understanding of the mechanisms which generate correlations and entanglement in large-scale atomic ensembles, particularly their robustness to noise, can also have useful applications in quantum technology. In particular, demonstrations of foundational tests may provide new ways to benchmark states for sub-shot-noise atom interferometry and applications in quantum information science.

In pursuit of answers to these questions, we have investigated and designed new proposals to demonstrate EPR entanglement and Bell inequality violations in systems of ultracold atomic gases. The results of this investigation are presented in this thesis, which can be broken into five main chapters.

In the remainder of this chapter we give an introduction to the essential physics relevant to the topics discussed in this thesis. Section 1.4 discusses the main systems and processes which we use to generate and understand entanglement and non-locality in this thesis. Specifically, we give a brief review of: (i) the general two-mode squeezed vacuum state and its role in demonstrating EPR entanglement, non-locality and other non-classical effects, (ii) spontaneous four-wave mixing via colliding Bose–Einstein condensates, including a theoretical background of the process and the correlations it generates, and (iii) spin-changing collisions in a spinor condensate, including a derivation of the basic Hamiltonian in the single-mode approximation.

Further to this, in Sect. 1.5 we give a simple introduction to phase-space representations of quantum mechanics, specifically the Wigner and positive P representations. In particular, we focus on the derivation and implementation of these representations with respect to numerically modelling physical systems. We also give illustrative examples in which the representations can be applied, specifically the processes of spontaneous four-wave mixing via colliding condensates and spin-changing collisions, which are the focus of this thesis.

Chapter 2 outlines a proposal to demonstrate the previously un-realized non-classical Hong–Ou–Mandel effect with massive particles, utilizing pair-correlated atoms produced by condensate collisions. This textbook effect of two-particle quantum interference serves not only as a demonstration of underlying non-classical correlations between the pairs of scattered atoms, but also as an important stepping-stone for experimental demonstration of a Bell inequality due to the similarity of the interferometric schemes we use in both.

Building on this, in Chap. 3 we outline a proposal to demonstrate a violation of a motional-state Bell inequality with massive particles via colliding condensates. Our scheme is an atom-optics analog of the Rarity–Tapster interferometer, which was previously used to demonstrate a violation using pairs of momentum-entangled photons [12]. However, key differences arise due to the multimode nature of the collision process. We use numerical simulations, based on stochastic phase-space methods, to calculate the necessary non-local correlations and explore the possibility of a Bell inequality violation via condensate collisions in experimentally relevant parameter regimes.

In Chap. 4 we investigate a recent experimental attempt to demonstrate continuous-variable EPR entanglement with a mesoscopic ensemble of massive particles, via spin-changing collisions in a spinor BEC [7]. An idealised theoretical analysis indicates that the state produced via these spin-changing collisions (the archetypal two-mode squeezed vacuum in the simplest approximation) should exhibit EPR entanglement, however, we investigate the possibility that the collisions are initiated by thermal fluctuations. We characterize the impact of these fluctuations on the dynamics of the system and how they may lead to the destruction of EPR entanglement.

Next, in Chap. 5 we investigate how the strong correlations of the two-mode squeezed vacuum state can be applied in the context of quantum metrology. We examine how a spinor BEC (identical to the experiment considered in Chap. 4) could be used to realize a quantum-enhanced 'active' atom interferometer, also known as a SU(1, 1) interferometer. Particularly, we outline how the phase-sensitive correlations between spinor components provide an ideal state for sub-shot-noise interferometry at the Heisenberg limit. Our theoretical analysis gives key results for an example spinor condensate in a realistic experimental regime. Furthermore, we also discuss how unique features of the atomic realization must be dealt with to realize the archetypal SU(1, 1) scheme [18] in a spinor BEC.

Lastly, in Chap. 6, building on the phase-space techniques used extensively throughout the previous chapters, we examine the interpretation of individual phase-space trajectories of the Wigner function as corresponding to possible outcomes of single experimental trials. To this end, we investigate the relation between the true (measured) particle number distribution P_n for a single-mode state and that obtained by discretely binning the individual stochastic realisations of squared mode amplitudes $|\alpha|^2$ of the sampled Wigner distribution $W(\alpha)$, which we denote via $\tilde{P}_n$. We find that there is indeed a close quantitative correspondence between P_n and $\tilde{P}_n$ for a wide range of states, justifying the broadly accepted view that, for highly occupied modes, individual stochastic realisations of Wigner trajectories should approximately correspond to outcomes of single experiments. However, we also find counterexamples for which high mode occupation may not be sufficient for such an interpretation; we find instead that a more relevant and sufficient requirement is the smoothness and broadness of the Wigner function $W(\alpha)$ for the state of interest relative to the scale of oscillations of the Wigner functions for the relevant Fock states.

1.4   Background I: Physical Systems

In our study of the generation and measurement of entanglement and non-classical correlations, we focus on two specific systems: spontaneous four-wave mixing via colliding condensates and spin-changing collisions in spinor condensates. In the simplest approximation, both of these pair-production process can be described by the archetypal two-mode squeezing Hamiltonian, which produces the textbook two-mode squeezed vacuum state prevalent in quantum optics. Such a state is known to possess strong non-classical correlations and has been the subject of intense study in terms of demonstrating EPR entanglement and quantum non-locality. Indeed, it is the relation to this state which motivates us to investigate spin-changing collisions and spontaneous four-wave mixing in the context of fundamental tests of quantum mechanics with massive particles.

The following chapter is composed of a brief summary and review of these topics which will form the basis of proposals in this thesis. In particular, we formally introduce the theoretical concept of the two-mode squeezed vacuum state and outline some essential results for the state in the context of non-classical correlations, EPR entanglement and non-locality. Building on this, we then briefly review the processes of spontaneous four-wave mixing and spin-changing collisions and outline their theoretical background, which will serve as a foundation for the work of Chaps. 2–5.

1.4.1   Entanglement and Non-locality with the Two-Mode Squeezed Vacuum State

The archetypal state which we focus upon heavily in this thesis is the two-mode squeezed vacuum state, defined as [19]

$$|\varphi\rangle = \hat{S}(\zeta)|0_1, 0_2\rangle, \tag{1.6}$$

where the two-mode squeezing operator is defined as $\hat{S}(\zeta) = e^{\zeta^* \hat{a}_1 \hat{a}_2 - \zeta \hat{a}_1^\dagger \hat{a}_2^\dagger}$ for $\zeta = re^{i\theta}$ and $\hat{a}_i$ ($\hat{a}_i^\dagger$) is the annihilation (creation) operator of the mode $i = 1, 2$. The squeezing operator acts on the vacuum state, denoted $|0_1, 0_2\rangle$, adding bosons pairwise into the modes 1 and 2.

A more useful representation of the state, in the Schrödinger picture, can be found by applying the SU(1, 1) disentangling theorem to the squeezing operator [19]. Arbitrarily setting $\theta = 0$, we may then rewrite Eq. (1.6) as a sum of twin-Fock states [6]:

$$|\varphi\rangle = \sqrt{1 - \lambda^2} \sum_{n=0}^{\infty} \lambda^n |n, n\rangle, \tag{1.7}$$

where $\lambda = \tanh(r)$. The two-mode squeezed vacuum state can be produced dynamically by the Hamiltonian,

$$\hat{H} = i\hbar\kappa \left(\hat{a}_1\hat{a}_2 - \hat{a}_1^\dagger\hat{a}_2^\dagger \right), \tag{1.8}$$

where $r = \kappa t$ for some real gain coefficient κ. Physically, this Hamiltonian is related to the production of bosons in pairs, populating modes $\hat{a}_1$ and $\hat{a}_2$ in a correlated manner. A common realization in quantum optics is by the processes of spontaneous parametric down-conversion, wherein a photon of frequency ω_0 passes through a $\chi^{(2)}$ nonlinear medium and splits into a pair of photons of frequencies ω_1 and ω_2 such that $\omega_0 = \omega_1 + \omega_2$. The simplest Hamiltonian describing this optical process is given by

$$\hat{H}_{DC} = i\hbar g \left(\hat{a}_0^\dagger\hat{a}_1\hat{a}_2 - \hat{a}_0\hat{a}_1^\dagger\hat{a}_2^\dagger \right), \tag{1.9}$$

where g is the gain which is related to the $\chi^{(2)}$ nonlinearity of the medium. Under usual experimental parameters, the $\hat{a}_0$ mode (referred to commonly as the pump) is a strong coherent field. Often, such a system is modeled by invoking the undepleted pump approximation, wherein the loss of photons in the pump due to the down-conversion process is negligible (in general, such that the population of the down-converted photons does not exceed $\sim 10\,\%$ of the pump mode) and one replaces the operator by a c-number, $\hat{a}_0 \simeq \alpha_0$, to give the relation to the squeezing Hamiltonian of Eq. (1.8), $\kappa \equiv g\alpha_0$, where we assume that α_0 is real without loss of generality.

As discussed in Sect. 1.1, this state has proven to be pivotal in demonstrations of non-classical correlations, entanglement and non-locality. This is not only due to its important physical properties but also due to the prevalence of processes wherein the underlying Hamiltonian may be reduced to Eq. (1.8). Such processes extend beyond the area of quantum optics into many physical systems, including the field of ultracold Bose gases. In particular, we utilize that the processes of spontaneous four-wave mixing of matter waves and spin-changing collisions of spinor condensates both produce, in the simplest approximation, the two-mode squeezed vacuum state in Chaps. 2–5.

Non-classical Correlations

The simplest feature of the two-mode squeezed state one can examine are the correlations between the $\hat{a}_1$ and $\hat{a}_2$ modes (commonly referred to as the signal and idler modes). From an inspection of Eq. 1.8, one expects due to the pair-wise nature of the production process that these modes will be strongly correlated. In particular, at the simplest level we expect that for every particle emitted into the signal mode there must be a corresponding partner emitted into the idler mode. To characterize these correlations in the system we introduce the second-order correlation function of Glauber [20],

$$g_{ij}^{(2)} = \frac{\langle : \hat{n}_i\hat{n}_j : \rangle}{\langle \hat{n}_i \rangle \langle \hat{n}_j \rangle}, \tag{1.10}$$

for $i, j = 1, 2$ where $\hat{n}_i = \hat{a}_i^\dagger\hat{a}_i$ is the particle number operator of mode $i = 1, 2$ and the colon notation indicates normal ordering of the relative creation and annihilation operators. The simplest interpretation of $g_{ij}^{(2)}$ is that it indicates the likelihood of a particle being detected in mode j given that one was detected in mode i. By dividing by the product of mean-occupations of modes i and j, one is effectively normalizing with respect to an uncorrelated Poissonian random process. A value of $g_{ij}^{(2)} > 1$ indicates that the modes are correlated, whilst $g_{ij}^{(2)} < 1$ is termed as anti-correlated. Evaluating $g_{ij}^{(2)}$ for the squeezed vacuum state we find the important results

$$g_{12}^{(2)} = 2 + \frac{1}{n}, \tag{1.11}$$

$$g_{11}^{(2)} = g_{22}^{(2)} = 2, \tag{1.12}$$

for $n \equiv \langle \hat{n}_1 \rangle = \langle \hat{n}_2 \rangle$. These correlations are commonly referred to as the cross- and auto-correlation respectively. The cross-correlation, $g_{12}^{(2)} > 1$ reflects the creation of pairs of particles in opposite modes, whilst the auto-correlation is a result of bosonic bunching analogous to the well known Hanbury–Brown–Twiss effect [21].

Correlations alone, however, are not in any sense a unique feature of quantum mechanics. Furthermore, the process of pair production is not deeply quantum in nature and has classical analogs. Specifically, one could construct some classical stochastic theory [22] wherein pairs are randomly generated in the signal and idler modes, such that the individual modes have classically fluctuating intensities, but will also be correlated with $g_{12}^{(2)} > 1$. However, the quantum nature of the correlations emerge when we consider the classical Cauchy–Schwarz inequality [23, 24],

$$g_{12}^{(2)} \leq \sqrt{g_{11}^{(2)} g_{22}^{(2)}}. \tag{1.13}$$

which places bounds on the relative strength of the cross-correlation with respect to the auto-correlation of each mode. The quantum theory of the two-mode squeezed vacuum violates this inequality for all n and thus the cross-correlation is stronger than classically allowed. Equivalently, this excludes any classical stochastic theory as a valid physical description of the pair-production process [22], as it will obey the Cauchy–Schwarz inequality and thus cannot replicate correlations as strong as those predicted by the quantum theory.

A consequence of this violation in this case is the presence of squeezing of the number difference between the signal and idler modes. It can be shown that the two-mode squeezed vacuum has vanishing fluctuations of the number difference, such that

$$\langle \Delta^2 \hat{n}_- \rangle = \langle \Delta^2 (\hat{n}_1 - \hat{n}_2) \rangle = 0. \tag{1.14}$$

where $\langle \Delta^2 \hat{A} \rangle = \langle \hat{A}^2 \rangle - \langle \hat{A} \rangle^2$ is the variance of the operator $\hat{A}$. Relative to the sum population of the side modes, $\langle \hat{n}_+ \rangle = \langle \hat{n}_1 + \hat{n}_2 \rangle$, we have that $\langle \Delta^2 \hat{n}_- \rangle < \langle \hat{n}_+ \rangle$ and the fluctuations are consequentially sub-Poissonian. This is an intuitively obvious result, as there exists no mechanism in the Hamiltonian of Eq. (1.8) to produce bosons individually. In contrast, one can show that there exist strong fluctuations in the sum population:

$$\langle \Delta^2 \hat{n}_+ \rangle = \langle \Delta^2 (\hat{n}_1 + \hat{n}_2) \rangle = \langle \hat{n}_+ \rangle \left(2 + \langle \hat{n}_+ \rangle\right), \tag{1.15}$$

which are super-Poissonian as $\langle \Delta^2 \hat{n}_+ \rangle > \langle \hat{n}_+ \rangle$. Notably, while the statistics of each individual mode are thermal, $\langle \Delta^2 \hat{n}_i \rangle = \langle \hat{n}_i \rangle (\langle \hat{n}_i \rangle + 1)$ for $i = 1, 2$, the combined fluctuations are stronger than one expects for a pair of un-correlated modes with thermal statistics, which would be equal to the sum of the individual variances.

EPR Entanglement

Non-classical correlations, in the sense of a Cauchy–Schwarz violation, are a prerequisite to demonstrate entanglement (in terms of both simple inseparability and the EPR paradox) and non-locality. However, the intensity–intensity correlations investigated so far are not sufficient to demonstrate EPR entanglement, instead we must show that there exist phase-sensitive correlations between the modes. To this end we introduce the quadrature operators

$$\hat{X}_j^\phi = \hat{a}_j e^{-i\phi} + \hat{a}_j^\dagger e^{i\phi}, \tag{1.16}$$

where $j = 1, 2$. In shorthand, one usually denotes $\hat{X}_j^0 \equiv \hat{X}_j$ and $\hat{X}_j^{\pi/2} \equiv \hat{Y}_j$. The $\hat{X}_j$ and $\hat{Y}_j$ quadratures are canonically conjugate operators with commutation relation $[\hat{X}_k, \hat{Y}_j] = 2i\delta_{kj}$ and are the quantum optics equivalent of position $\hat{x}$ and momentum $\hat{p}$ operators for particles with non-zero rest mass, as originally considered by EPR. In practice, such operators are measured via homodyne detection, wherein the signal or idler mode is mixed with a strong coherent field on a 50–50 beam-splitter, i.e. a laser in the case of down-converted photons. After mixing, a measurement of the relative number difference of the output ports of the beam-splitter is proportional to the quadrature amplitudes.

To demonstrate EPR entanglement for the two-mode squeezed vacuum state, one can readily follow the prescription of Reid [5] outlined in Sect. 1.1. Firstly, we can demonstrate that there exist correlations between the quadratures,

$$\langle \hat{X}_1^\phi \hat{X}_2^\vartheta \rangle = 2\sinh(r)\cosh(r)\sin(\phi + \vartheta), \tag{1.17}$$

such that for large $r \gg 1$ the cross-correlation maximally saturates the Cauchy–Schwarz inequality for $\phi + \vartheta = 0$, e.g.,

$$\langle \hat{X}_1 \hat{X}_2 \rangle = \sqrt{\langle (\hat{X}_1)^2 \rangle \langle (\hat{X}_2)^2 \rangle}. \tag{1.18}$$

and similarly for $\hat{Y}_1$ and $\hat{Y}_2$. Following the EPR argument, this strong correlation between quadratures implies that a measurement of, e.g., $\hat{X}_1$ would allow one to 'infer' the outcome of a measurement of $\hat{X}_2$. The error in this inference will be dependent on the degree of correlation between the quadratures and, using a linear inference method, the minimum error is given by [5],

$$\Delta_{\text{inf}}^2 \hat{X}_j \equiv \langle \Delta^2 \hat{X}_j \rangle - \frac{\langle \Delta \hat{X}_k \Delta \hat{X}_j \rangle}{\langle \Delta^2 \hat{X}_k \rangle}, \tag{1.19}$$

and similarly for $\hat{Y}_j$. Using the commutation relations of the quadrature operators, $[\hat{X}_j, \hat{Y}_j] = 2i$, one can then construct the Heisenberg uncertainty relation for the product of quadrature variances,

$$\Delta^2 \hat{X}_1 \Delta^2 \hat{Y}_1 \geq 1, \tag{1.20}$$

which is seemingly violated by the *inferred* quadrature variances in the case of $r > 0$,

$$\Delta_{\text{inf}}^2 \hat{X}_1 \Delta_{\text{inf}}^2 \hat{Y}_1 = \frac{1}{\cosh^2(2r)} = \frac{1}{2n+1} < 1. \tag{1.21}$$

We take such a seeming violation to imply entanglement of the signal and idler mode in the sense of the EPR paradox. In the interpretation of EPR, the inferred value (and thus the respective uncertainty) was equivalent to the pre-existing *element of reality* and thus such a violation was in contradiction with quantum mechanics. However, in the standard interpretation of quantum mechanics, measurements of $\hat{X}_2$ and $\hat{Y}_2$ (and thus the respective inferrance of $\hat{X}_1$ and $\hat{Y}_1$) would need to be made in seperate experimental trials and hence the inferred quantities would not exist simultaneously.

An important observation regarding this result is that the inequality is always violated for $n > 0$ and the degree of violation increases as $\sim 1/2n$ for $n \gg 1$. Consequentially, one expects that for a two-mode squeezed vacuum state in an ideal experiment it is possible to demonstrate EPR entanglement between two modes that may be macroscopically populated.

Demonstrating Non-locality via a Bell Inequality

Further to the demonstration of EPR entanglement, one may use a pair of two-mode squeezed vacuum states [in, e.g., modes (1, 2) and (3, 4)] to demonstrate non-locality in the sense of a Bell inequality violation. However, unlike the previous discussion of EPR entanglement, we demonstrate that this violation can only be demonstrated in specific regimes. In particular, a pair of two-mode squeezed states in the weak-gain regime can be used to form an approximation to an ideal Bell state, which violates the inequality maximally.

We focus our discussion on the Rarity–Tapster scheme as this is the most pertinent to this thesis, particularly as it forms the basis to the experimental proposal of Chap. 3. The Rarity–Tapster scheme, illustrated in Fig. 1.2, utilizes a pair of two-mode squeezed states as the input to a four-mode interferometer. This input state can be written as a product of the two squeezed states

$$|\Psi\rangle = (1 - \lambda^2) \sum_{k,m=0}^{\infty} \lambda^{(k+m)} |k\rangle_1 |k\rangle_2 |m\rangle_3 |m\rangle_4. \tag{1.22}$$

where $\lambda = \tanh(r)$ and the subscript denotes the modes $i = 1, 2, 3, 4$. The state may be generated by a four-mode generalization of the squeezing Hamiltonian [Eq. (1.8)],

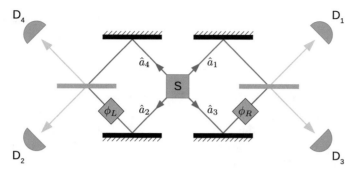

Fig. 1.2 Rarity–Tapster interferometer, with a multimode source S which produces a pair of two-mode squeezed vacuum states. Experimentally, this interferometer can be realized by using two pairs of photons produced by spontaneous optical parametric down-conversion [12]. After passing through the interferometer, particles can be detected in the upper/lower paths at the output at detectors D_i for $i = 1, 2, 3, 4$

$$\hat{H} = i\hbar\kappa \left(\hat{a}_1 \hat{a}_2 + \hat{a}_3 \hat{a}_4 - \hat{a}_1^\dagger \hat{a}_2^\dagger - \hat{a}_3^\dagger \hat{a}_4^\dagger \right), \tag{1.23}$$

where $r = \kappa t$ as previously.

By including a phase-shift (ϕ_L and ϕ_R) in the two lower arms of the interferometer (with the phase-shifts being relative to the upper arms) and mixing the pairs of modes from each squeezed state on separate beam-splitters, one is able to demonstrate there exist phase-dependent intensity–intensity correlations between the different output ports of the interferometer, corresponding to correlation measurements between the detectors (see Fig. 1.2),

$$C_{12} = C_{34} = n^2 + \frac{n(1+n)}{2} \left[1 - \cos(\phi_R - \phi_L) \right], \tag{1.24}$$

$$C_{14} = C_{23} = n^2 + \frac{n(1+n)}{2} \left[1 + \cos(\phi_R - \phi_L) \right], \tag{1.25}$$

where

$$C_{ij} \equiv \langle \hat{n}_i \hat{n}_j \rangle, \tag{1.26}$$

and $n \equiv \langle \hat{a}_i^\dagger \hat{a}_i \rangle = \sinh^2(r)$ is the occupation of the individual modes $i = 1, 2, 3, 4$ of the squeezed states as previously defined. The CHSH-Bell quantity S can then be constructed from these correlations and, by an appropriate choice of four pairs of phase-settings (see Chap. 3 for further details), we find the result

$$S = 2\sqrt{2}\frac{1+n}{1+3n}. \tag{1.27}$$

As outlined in Sect. 1.1, any local hidden-variable theory is bounded by the inequality $S \leq 2$, whereas for small $n \to 0$ we have $S \to 2\sqrt{2}$, which is the maximal violation predicted by quantum mechanics. The dependence of S on occupation n is an interesting result for the two-mode squeezed vacuum state, as it places an upper bound of $n < (\sqrt{2} - 1)/(3 - \sqrt{2}) \simeq 0.26$ for $S > 2$. Thus, unlike for demonstration of the EPR paradox, a Bell inequality violation decreases as the population of the modes increases.

The fundamental reason for this scaling can be understood by examining the form of the input state to the interferometer. In the weak-gain regime of the two-mode squeezed vacuum, which corresponds to $\lambda \simeq r \ll 1$ and hence an average mode occupation in each of the four modes of $n \simeq \lambda^2 \simeq r^2 \ll 1$, the sum over Fock states in Eq. (1.22) can be truncated to lowest order in λ,

$$\begin{aligned} |\Psi\rangle &\propto |0\rangle_1 |0\rangle_2 |0\rangle_3 |0\rangle_4 \\ &+ \lambda(|1\rangle_1 |1\rangle_2 |0\rangle_3 |0\rangle_4 + |0\rangle_1 |0\rangle_2 |1\rangle_3 |1\rangle_4). \end{aligned} \tag{1.28}$$

Taking into account the fact that the contribution from the pure vacuum state (the first term) does not affect the outcome of any correlation measurements, we can further approximate this state by $|\Psi\rangle \propto \alpha(|1\rangle_1 |1\rangle_2 |0\rangle_3 |0\rangle_4 + |0\rangle_1 |0\rangle_2 |1\rangle_3 |1\rangle_4)$, which can be mapped to the archetypal Bell state $|\Psi^+\rangle = \frac{1}{\sqrt{2}}(|+\rangle_L |-\rangle_R + |-\rangle_L |+\rangle_R)$ in the polarization or spin-$1/2$ $\hat{S}_z$ basis, where the subscript (L,R) refers to the left and right arms of the interferometer. This ideal Bell state is known to maximally violate the CHSH inequality with $S = 2\sqrt{2}$.

The scaling of the violation of the Bell inequality can thus be understood purely in terms of contributions of higher-order Fock states to Eq. (1.22), which leads to a breakdown of the mapping to the ideal Bell state of Eq. (3.1). This is an important result to appreciate, as the two-mode squeezed vacuum is commonly used in the weak-gain regime in quantum optics to produce an effective twin-photon state $|\varphi\rangle \sim |1_1 1_2\rangle$, for instance in the Hong–Ou–Mandel effect, whose matter-wave analog is studied in Chap. 2.

Overall, these results show that the two-mode squeezed vacuum is a versatile and important state in foundational tests of quantum mechanics. Building upon this, by mapping simple dynamical process in ultracold gases back to the fundamental Hamiltonian of Eq. (1.8) we can identify systems which may be candidates to generate entanglement and non-local correlations with massive particles. In this light, the generic results we have derived for the ideal two-mode squeezed vacuum state are fundamental to the results of Chaps. 2–5.

1.4.2 Spontaneous Four-Wave Mixing in Colliding Bose–Einstein Condensates

Spontaneous four-wave mixing of matter waves has been a topic of strong interest in recent years in the atom-optics community. Experimentally, pioneering work has been performed by the group of Aspect and Westbrook. Colliding a pair of metastable

Helium (^{4}He*) BECs, they generated a *collision halo* of atoms due to s-wave scattering [25] and measured the resulting atom-atom pair correlations. In the simplest approximation, the underlying process which generates the collision halo can be shown to be equivalent to the squeezing Hamiltonian of Eq. (1.8) which produces the two-mode squeezed vacuum state, albeit generalized to a multimode form. Building upon the results of Sect. 1.4.1, we thus expect that condensates collisions should be a promising candidate to investigate entanglement and non-locality in a system of massive particles.

In this vein, subsequent experimental and theoretical work has enabled the measurement and characterization of second-order coherence and strong correlations between atoms occupying opposing regions of the halo [26, 27]. These correlations were shown to violate the classical Cauchy–Schwarz inequality [28] and led to a measurement of a sub-Poissonian number difference [29]. In the following we give a brief review of these pioneering results, which have paved the way for foundational tests of quantum mechanics such as demonstrating the Hong–Ou–Mandel effect (see Chap. 2) and violation of a Bell inequality (see Chap. 3) in ultracold gases.

In the experiments of Aspect and Westbrook, approximately 1.5×10^5 atoms are initially held in an anisotropic magnetic trap, leading to an elongated (cigar-shaped) BEC. Applying a Raman transition (and simultaneously turning off the trap), the BEC is split into two counter-propagating halves with momenta $\mathbf{k}_1 = -\mathbf{k}_2$ and $|\mathbf{k}_1| = |\mathbf{k}_2| = k_0$. As the condensates spatially separate, constituent atoms of each undergo s-wave scattering. In particular, atoms with momenta $\mathbf{k}_1$ and $\mathbf{k}_2$ may collide and create a new pair of atoms with momenta $\mathbf{k}_3$ and $\mathbf{k}_4$ satisfying momentum conservation such that $\mathbf{k}_1 + \mathbf{k}_2 = \mathbf{k}_3 + \mathbf{k}_4 = 0$ and therefore $\mathbf{k}_3 = -\mathbf{k}_4$. Further, energy conservation implies that $|\mathbf{k}_3| = |\mathbf{k}_4| = k_0$, i.e. the atoms are scattered into two equal but opposite momentum modes situated on the surface of a sphere of radius k_0 in momentum space as illustrated in Fig. 1.3a. In typical experiments approximately 5 % of the initial condensate population is scattered into the collision halo, which can be imaged in position-space using a micro-channel plate (MCP) detector after ballistic time-of-flight expansion, illustrated in Fig. 1.3b. The excellent temporal and spatial resolution of the detector allows single-atom detection and enables experimentalists to fully re-construct the 3D momentum-space distribution of the scattered atoms from time-of-flight and position data. A typical experimental cross-section of the collision halo and comparison with theoretical simulations of Ref. [26] is shown in Fig. 1.4.

Extensive work has been carried out to theoretically and experimentally characterize the process of spontaneous four-wave mixing in colliding condensates. Importantly for the results of Chaps. 2 and 3, there has been shown to be excellent qualitative and quantitative agreement between first-principles numerical simulations and experimental observations on many features of the collision halo, including variations in the radius and thickness of the halo [31, 32] which are not encompassed by simpler analytic models [33] (see Appendix A for further details).

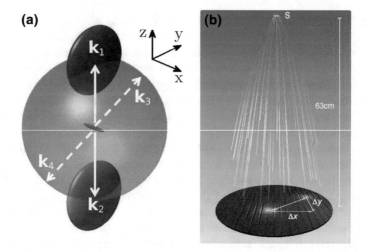

Fig. 1.3 a Schematic of the four-wave mixing scheme in momentum-space. The counter-propagating halves of the condensate are represented by the pancakes (cigar shaped in reciprocal space—pictured in the *center*) at $\mathbf{k}_{1,2}$. Constituent atoms of the two halves collide and scatter into new modes $\mathbf{k}_3$ and $\mathbf{k}_4$ satisfying energy and momentum conservation such that $\mathbf{k}_1 + \mathbf{k}_2 = \mathbf{k}_3 + \mathbf{k}_4$. These states lie on a spherical shell of radius k_0 known as the collision halo. **b** Illustration of the experimental detection process. Atoms are scattered from the condensate into the collision halo. The cloud of atoms expands freely in position-space and falls onto a MCP detector. The 3D momentum distribution can then be re-constructed from recorded time-of-flight and position data. Figure adopted from Ref. [30]

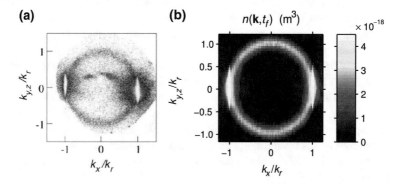

Fig. 1.4 a Typical experimental cross-section of collision halo (after averaging over experimental trials) and **b** numerical simulation. The condensate is evidenced by the highly populated (*white*) regions along the k_x-axis (the geometry here differs to the example of Fig. 1.3) whilst the ring of the collision halo is clearly visible. Figures are adopted from Refs. [25, 26]

The Hamiltonian of the scattering process can be written simply as

$$\hat{H} = \int d^3\mathbf{r}\, \hat{\psi}^\dagger(\mathbf{r}) \left[-\frac{\hbar^2}{2m}\nabla^2 \right] \hat{\psi}(\mathbf{r}) + \frac{U}{2}\hat{\psi}^\dagger(\mathbf{r})\hat{\psi}^\dagger(\mathbf{r})\hat{\psi}(\mathbf{r})\hat{\psi}(\mathbf{r}), \qquad (1.29)$$

where $U = 4\pi\hbar^2 a/m$ is the interaction strength characterized by the s-wave scattering length a between atoms described by the creation (annihilation) field operators $\hat{\psi}^\dagger$ ($\hat{\psi}$). The collision is dominated by the interaction term of the Hamiltonian, however, due to the quartic nature of the Hamiltonian it is not exactly solvable. A positive P-representation of the full field $\psi(\mathbf{r}, t)$ allows one to numerically model the dynamics, however, this is usually restricted to time-scales shorter than the total collision duration. In general, only a small fraction of the initial condensate population is scattered into the collision halo and thus the system lends itself to an analysis using a linearization procedure. We invoke a Bogoliubov approximation to split the full field operator $\hat{\psi} = \psi + \hat{\delta}$ into mean-field ψ and fluctuating $\hat{\delta}$ components. An effective Hamiltonian governing the dynamics of the fluctuating component can then be found [34, 35],

$$
\begin{aligned}
\hat{H}_{\text{eff}} = \int d^3\mathbf{r}\, \Big\{ & \hat{\delta}^\dagger(\mathbf{r}, t) \left[-\frac{\hbar^2}{2m}\nabla^2 \right] \hat{\delta}(\mathbf{r}, t) + 2U\, |\psi(\mathbf{r}, t)|^2\, \hat{\delta}^\dagger(\mathbf{r}, t)\hat{\delta}(\mathbf{r}, t) \\
& + U \Big[\psi_{+\mathbf{k}_0}(\mathbf{r}, t)\psi_{-\mathbf{k}_0}(\mathbf{r}, t)\hat{\delta}^\dagger(\mathbf{r}, t)\hat{\delta}^\dagger(\mathbf{r}, t) \\
& + \psi^*_{+\mathbf{k}_0}(\mathbf{r}, t)\psi^*_{-\mathbf{k}_0}(\mathbf{r}, t)\hat{\delta}(\mathbf{r}, t)\hat{\delta}(\mathbf{r}, t) \Big] \Big\},
\end{aligned}
\qquad (1.30)
$$

where we have further reduced the mean-field wavefunction into the two counter-propagating components $\psi(\mathbf{r}, t) = \psi_{+\mathbf{k}_0}(\mathbf{r}, t) + \psi_{-\mathbf{k}_0}(\mathbf{r}, t)$.

The bilinear form of the interaction term in the second and third lines of Eq. (1.30) can be recognized as a generalized (multimode) form of Eq. (1.8) once rewritten in momentum space via a Fourier transform, where the effective (inhomogeneous) nonlinearity $\kappa(\mathbf{r}, t) \propto (U/2)\psi_{+\mathbf{k}_0}(\mathbf{r}, t)\psi_{-\mathbf{k}_0}(\mathbf{r}, t)$ is provided by the coupling U and the product of the mean-field amplitudes of the colliding condensates. In momentum space, this interaction term corresponds to pair-production of atoms into modes $\pm\mathbf{k}$ approximately situated on the spherical halo of radius k_0. When the fluctuating component $\hat{\delta}$ is taken to be initially in the vacuum state and considering for simplicity a homogeneous system in a finite box (hence discretized momentum modes) where we invoke the undepleted pump approximation such that the number of atoms in the condensate is taken to be fixed, we expect the atom pairs scattered into the collision halo to be well approximated by a multimode product state of two-mode squeezed vacuum states which has the form,

$$|\psi_{4WM}\rangle \equiv \bigotimes_{\mathbf{k}>0} |\psi_{\mathbf{k}}\rangle, \qquad (1.31)$$

for

$$|\psi_{\mathbf{k}}\rangle = \sqrt{1 - \lambda_{\mathbf{k}}^2} \sum_{n=0}^{\infty} \lambda_{\mathbf{k}}^n |n_{-\mathbf{k}}, n_{\mathbf{k}}\rangle, \tag{1.32}$$

where $\lambda_{\mathbf{k}}$ will be determined by the nonlinearity $\kappa(\mathbf{r}, t)$ and collision duration t. As highlighted in Sect. 1.4.1, such a state has strong correlations between output modes, which in this case corresponds to correlations between atoms in modes $\pm \mathbf{k}$.

We can generalize the discrete second-order correlation $g_{ij}^{(2)}$ introduced in Sect. 1.4.1 to the continuous momentum-space second-order correlation function

$$g^{(2)}(\mathbf{k}, \mathbf{k}') = \frac{\langle \hat{a}^\dagger(\mathbf{k}) \hat{a}^\dagger(\mathbf{k}') \hat{a}(\mathbf{k}') \hat{a}(\mathbf{k}) \rangle}{\langle \hat{a}^\dagger(\mathbf{k}) \hat{a}(\mathbf{k}) \rangle \langle \hat{a}^\dagger(\mathbf{k}') \hat{a}(\mathbf{k}') \rangle}, \tag{1.33}$$

where $\hat{a}^\dagger(\mathbf{k})$ [$\hat{a}(\mathbf{k})$] is the Fourier transform of $\hat{\delta}^\dagger(\mathbf{r}, t)$ [$\hat{\delta}(\mathbf{r}, t)$], or more descriptively the creation (annihilation) operator of an atom scattered into the collision halo with momentum $\mathbf{k}$. Physically, $g^{(2)}(\mathbf{k}, \mathbf{k}')$ can be interpreted identically to the discrete form $g_{ij}^{(2)}$, corresponding to likelihood of detection of an atom with momentum $\mathbf{k}$ given that another with momentum $\mathbf{k}'$ has been detected. Again, a level of $g^{(2)}(\mathbf{k}, \mathbf{k}') > 1$ indicates an increased likelihood of this event and $g^{(2)}(\mathbf{k}, \mathbf{k}') < 1$ a decreased likelihood, whilst $g^{(2)}(\mathbf{k}, \mathbf{k}') = 1$ indicates an uncorrelated process with respect to random events distributed according to Poissonian statistics.

Motivated by the results for the idealized two-mode squeezed vacuum state, we can identify two important cases of $g^{(2)}(\mathbf{k}, \mathbf{k}')$ in the collision halo. Firstly, we identify a 'collinear' (CL) correlation for $\mathbf{k}' = \mathbf{k}$ [equivalent to $g_{11}^{(2)}$ and $g_{22}^{(2)}$ in the simple model of Sect. 1.4.1]. Secondly, we characterize a 'back-to-back' (BB) correlation for scattered atom pairs such that $\mathbf{k} = -\mathbf{k}$ [equivalent to $g_{12}^{(2)}$ in the simple model of Sect. 1.4.1]. To give illustrative results, we can use the simplest model of the scattering process, corresponding to two BECs of uniform density colliding in a finite quantization volume (see Appendix A for details of the calculation). The correlations between discretized momentum modes in this case reduce to

$$g_{\mathbf{k}, \mathbf{k}}^{(2)} = 2, \tag{1.34a}$$

$$g_{\mathbf{k}, -\mathbf{k}}^{(2)} = 2 + \frac{1}{n_{\mathbf{k}}}, \tag{1.34b}$$

where $n_{\mathbf{k}} = \langle \hat{a}_{\mathbf{k}}^\dagger \hat{a}_{\mathbf{k}} \rangle$ is the occupation of momentum mode $\mathbf{k}$ and we assume $n_{\mathbf{k}} = n_{-\mathbf{k}}$. The collinear correlation indicates the Hanbury–Brown–Twiss (HBT) effect of bosonic bunching whilst the strong back-to-back correlation is due to momentum and energy conservation in the s-wave scattering process. As the atoms scatter in pairs, it is intuitive that under these constraints the detection of one atom with momentum $\mathbf{k}$ will be accompanied by the detection of another atom with momentum $-\mathbf{k}$ from the same scattering event.

To demonstrate the non-classicality of these correlations, we can use the Cauchy–Schwarz inequality in the form

$$g^{(2)}_{\mathbf{k},-\mathbf{k}} \leq \sqrt{g^{(2)}_{\mathbf{k},\mathbf{k}} g^{(2)}_{-\mathbf{k},-\mathbf{k}}}. \tag{1.35}$$

Clearly, for the simple results of Eqs. (1.34a) and (1.34b) this inequality is violated for all $n_{\mathbf{k}}$. Such a violation was also experimentally demonstrated in Ref. [28] (although for a slightly more generalized multimode form of the inequality). As previously highlighted, a violation of the inequality is a pre-requisite for other non-classical phenomena of quantum mechanics such as demonstration of the Hong–Ou–Mandel effect and violation of a Bell inequality and thus gives strong support to the viability of such experiments in this system.

Although for illustrative purposes we have presented simple results here using a homogeneous BEC, experimentally one deals with inhomogeneous condensates with a momentum and position-space density distribution of some characteristic width. In this case, unlike the results of the ideal two-mode squeezed vacuum state, the correlation function $g^{(2)}(\mathbf{k}, \mathbf{k}')$ will have a finite correlation length around the peak-values at $\mathbf{k}' = \mathbf{k}$ and $\mathbf{k}' = -\mathbf{k}$. This can be illustrated by defining the second-order correlation functions

$$g^{(2)}_{\mathrm{CL}}(\mathbf{k}, \mathbf{k} + \mathbf{e}_i \Delta k_i) = \frac{\langle \hat{a}^\dagger(\mathbf{k}) \hat{a}^\dagger(\mathbf{k} + \mathbf{e}_i \Delta k_i) \hat{a}(\mathbf{k} + \mathbf{e}_i \Delta k_i) \hat{a}(\mathbf{k}) \rangle}{\langle \hat{a}^\dagger(\mathbf{k}) \hat{a}(\mathbf{k}) \rangle \langle \hat{a}^\dagger(\mathbf{k} + \mathbf{e}_i \Delta k_i) \hat{a}(\mathbf{k} + \mathbf{e}_i \Delta k_i) \rangle}, \tag{1.36}$$

$$g^{(2)}_{\mathrm{BB}}(\mathbf{k}, -\mathbf{k} + \mathbf{e}_i \Delta k_i) = \frac{\langle \hat{a}^\dagger(\mathbf{k}) \hat{a}^\dagger(-\mathbf{k} + \mathbf{e}_i \Delta k_i) \hat{a}(-\mathbf{k} + \mathbf{e}_i \Delta k_i) \hat{a}(\mathbf{k}) \rangle}{\langle \hat{a}^\dagger(\mathbf{k}) \hat{a}(\mathbf{k}) \rangle \langle \hat{a}^\dagger(-\mathbf{k} + \mathbf{e}_i \Delta k_i) \hat{a}(-\mathbf{k} + \mathbf{e}_i \Delta k_i) \rangle}, \tag{1.37}$$

where Δk_i is a displacement and $\mathbf{e}_i$ is the respective unit vector along the $i = x, y, z$ directions. Experimental and theoretical results [26, 34] have shown that these functions can readily be approximated by a Gaussian form,

$$g^{(2)}_{\mathrm{CL}}(\mathbf{k}, \mathbf{k} + \mathbf{e}_i \Delta k_i) \equiv 1 + h_{\mathrm{CL}} \prod_{i=x,y,z} e^{-\Delta k_i / 2\sigma_{\mathrm{CL},i}}, \tag{1.38}$$

$$g^{(2)}_{\mathrm{BB}}(\mathbf{k}, \mathbf{k} + \mathbf{e}_i \Delta k_i) \equiv 1 + h_{\mathrm{BB}} \prod_{i=x,y,z} e^{-\Delta k_i / 2\sigma_{\mathrm{BB},i}}, \tag{1.39}$$

where $h_{\mathrm{CL(BB)}}$ represents the height of the correlation above the background level of unity and $\sigma_{\mathrm{CL(BB)},i}$ is the rms correlation width along the $i = x, y, z$ direction. Typical values of the correlation lengths are on the order of the rms width of the momentum distribution of the condensate $\sigma_{\mathrm{CL(BB)},i} \sim \sigma_i$, with a pre-factor determined by the initial condensate density profile (see, e.g., Refs. [26, 34] and Appendix A for examples). Experimental characterization of these length-scales have proven to be crucial to understanding the correlations within the collision halo. Specifically, analysis of the correlation lengths led to a realization of a multimode Cauchy–Schwarz inequality violation [28]. They will also be important parameters in demonstration of the HOM effect and violations of a Bell inequality (see Chaps. 2 and 3 for more detail).

1.4.3 Spinor BEC

Spinor Bose–Einstein condensates are an increasingly popular area of study in the field of ultracold atomic gases. This popularity has been fueled by the interesting physics which emerges due to the addition of an internal (spin) degree of freedom to the usual spatial degrees of freedom, including the study of spin textures [36] and symmetry breaking [37]. Beyond mean-field analysis of the ground state of these systems there has also been extensive interest in coherent dynamics involving collisions between magnetic substates of the spinor BEC. Of particular interest is the collision process for a spin-1 (and also spin-2) condensate where two atoms in the $m_F = 0$ substate collide to produce a pair of atoms in the $m_F = \pm 1$ substate respectively. As we will demonstrate, such a pair-production process can be reduced, in the simplest approximation, to the squeezing Hamiltonian of Eq. (1.8) and thus we expect the idealized output state of the $m_F = \pm 1$ pair to be in the two-mode squeezed vacuum state. The entanglement and non-classical correlations present in this state make spin-changing collisions an obvious candidate for fundamental tests of quantum mechanics. In particular, unlike the condensate collision scheme outlined in the previous section which involves many populated modes in the collision halo, we consider small spinor BECs [e.g., of small atom number $\sim$150–500, and tightly confined by a harmonic potential] such that the spatial dynamics are frozen and we can reduce the problem to only a few modes, corresponding to the internal degrees of freedom. This simplification allows us to deal with mesoscopic ensembles of atoms which provide an ideal platform for both a previously unrealized demonstration of EPR entanglement with massive particles [38] and applications in quantum metrology.

The process of spin-changing collisions has been studied extensively theoretically [39–42] and demonstrated in various experimental systems [43–47]. Further experiments have verified the presence of entanglement between the output $m_F = \pm 1$ states [38] while recent work has also focused on the reversible nature of the process [48–50].

Here we outline the basic theoretical background of spinor BECs, in particular, we derive the general Hamiltonian which governs spatial and spin dynamics in Sect. 1.4.3 before demonstrating how the system can be reduced to a squeezing Hamiltonian for the spin degrees of freedom in Sect. 1.4.3. This theory will be built on further in Chaps. 4 and 5, wherein we investigate the feasibility of using spin-changing collisions to demonstrate EPR entanglement between mesoscopic ensembles of massive particles and a realization an atomic SU(1, 1) interferometer.

Basic Hamiltonian of a Spinor BEC

The general Hamiltonian of a spinor BEC can be broken into two parts [51, 52],

$$\hat{H} = \hat{H}_0 + \hat{H}_{\text{int}}, \tag{1.40}$$

where

$$\hat{H}_0 = \int d^3\mathbf{r} \sum_{m=-F}^{F} \hat{\psi}_m^\dagger(\mathbf{r}) \left[-\frac{\hbar^2}{2M}\nabla^2 + V(\mathbf{r}) + m\hbar p + |m|\hbar q \right] \hat{\psi}_m(\mathbf{r}), \quad (1.41)$$

is the usual single-particle Hamiltonian of kinetic energy and potential terms for a particle of mass M, as well as incorporating the linear and quadratic Zeeman shifts and $\hat{\psi}_m(\mathbf{r})$ [$\hat{\psi}_m^\dagger(\mathbf{r})$] is the bosonic annihilation (creation) operator of the $m = -F, -F+1, \ldots, F$ magnetic (m_F) substate of the atomic field in the hyperfine state F. The linear Zeeman shift is parameterized by $p = g\mu_B B/\hbar$ where g is the Landé hyperfine g-factor, μ_B the Bohr magneton and B the external magnetic field. Similarly, the quadratic shift is given by $q = \hbar p^2/\Delta E_{\mathrm{hf}}$ where ΔE_{hf} is the hyperfine energy splitting. The form of $\hat{H}_{\mathrm{int}}$ is non-trivial and we follow the derivation of Refs. [51, 52].

In this derivation we assume that the BEC is sufficiently dilute that we only need to consider binary interactions. Collisions between atoms can then be considered as scattering events between incoming and outgoing states characterized by total orbital angular momentum $\mathcal{L}_{\mathrm{pair}}$ and magnetic quantum number $m_{\mathcal{L},\mathrm{pair}}$. This interaction can be split into short and long-range components, the latter of which involves dipolar interactions and we neglect them here for simplicity. The short-range interaction is given by the conventional delta-potential form and is characterized by a length scale r_0, which we assume is much shorter than the de Broglie wavelength $\lambda_{\mathrm{dB}} \gg r_0$ (known as the cold-collision approximation [52]). This assumption has the consequence that only the lowest-order partial waves undergo interactions, thus $\mathcal{L}_{\mathrm{pair},i} = 0$ where the subscript indicates i the incoming (initial) state. Next, we assume that the interaction potential is rotationally invariant, such that the sum of the orbital $\hat{\mathcal{L}}_{\mathrm{pair}}$ and internal $\hat{\mathcal{F}}_{\mathrm{pair}}$ angular momentum of the colliding pair is conserved and, furthermore, by making the 'weak-dipolar approximation' [52] we neglect spin-orbit coupling and thus $\hat{\mathcal{L}}_{\mathrm{pair}}$ and $\hat{\mathcal{F}}_{\mathrm{pair}}$ are each also separately conserved, giving $\mathcal{L}_{\mathrm{pair},i} = \mathcal{L}_{\mathrm{pair},f} = 0$ and $\mathcal{F}_{\mathrm{pair},i} = \mathcal{F}_{\mathrm{pair},f}$ where the subscript f indicates the final state. Finally, we neglect hyperfine relaxation in our treatment and thus consider only collisions between atoms in the same hyperfine state F.

Following these approximations we may characterize the interactions by occurring in discrete spin channels, giving the general form of the interaction Hamiltonian

$$\hat{H}_{\mathrm{int}} = \sum_{\mathcal{F}_{\mathrm{pair}}=0,2,\ldots}^{2F} \hat{V}^{\mathcal{F}_{\mathrm{pair}}}, \quad (1.42)$$

where the summation is only over even $\mathcal{F}_{\mathrm{pair}}$ due to the symmetry of the wavefunction and

$$\hat{V}^{\mathcal{F}_{\mathrm{pair}}} = \frac{1}{2} g_{\mathcal{F}_{\mathrm{pair}}} \int d^3\mathbf{r} \, |\mathcal{F}_{\mathrm{pair}}, \mathcal{M}_{\mathrm{pair}}\rangle\langle\mathcal{F}_{\mathrm{pair}}, \mathcal{M}_{\mathrm{pair}}|, \quad (1.43)$$

with $|\mathcal{F}_{\text{pair}}, \mathcal{M}_{\text{pair}}\rangle$ a two-body state characterized by total angular momentum $\mathcal{F}_{\text{pair}}$ and total magnetic quantum number $\mathcal{M}_{\text{pair}}$. The coupling strength in the spin channel is defined as $g_{\mathcal{F}_{\text{pair}}} = 4\pi\hbar^2 a_{\mathcal{F}_{\text{pair}}}/M$ where $a_{\mathcal{F}_{\text{pair}}}$ is the s-wave scattering length.

By projecting onto single-body states we transform $\hat{V}^{\mathcal{F}_{\text{pair}}}$ to

$$\hat{V}^{\mathcal{F}} = \frac{1}{2}g_{\mathcal{F}} \int d^3\mathbf{r} \sum_{\mathcal{M}=-\mathcal{F}}^{\mathcal{F}} \hat{A}_{\mathcal{F}\mathcal{M}}^{\dagger}(\mathbf{r}, \mathbf{r})\hat{A}_{\mathcal{F}\mathcal{M}}(\mathbf{r}, \mathbf{r}), \qquad (1.44)$$

where we have adopted the shorthand $\mathcal{F} = \mathcal{F}_{\text{pair}}$ and $\mathcal{M} = \mathcal{M}_{\text{pair}}$ for brevity. The operator $\hat{A}_{\mathcal{F}\mathcal{M}}(\mathbf{r}, \mathbf{r}')$ is an irreducible operator which corresponds to annihilating a pair of bosons at $\mathbf{r}$ and $\mathbf{r}'$

$$\hat{A}_{\mathcal{F}\mathcal{M}}(\mathbf{r}, \mathbf{r}') = \sum_{m,m'=-F}^{F} \langle \mathcal{F}\mathcal{M}|Fm; Fm'\rangle \hat{\psi}_m(\mathbf{r})\hat{\psi}_{m'}(\mathbf{r}'), \qquad (1.45)$$

where $\hat{\psi}_m(\mathbf{r})$ is the atomic field operator of the $m = -F \dots F$ magnetic substate and $\langle \mathcal{F}\mathcal{M}|Fm; Fm'\rangle$ are Clebsch–Gordan coefficients due to the projection onto single-body states.

For the case of $F = 1$, calculation of the required Clebsch–Gordan coefficients [51] and subsequent substitution of Eq. (1.45) into Eq. (1.42) leads to the interaction Hamiltonian in the compact form [51]

$$\hat{H}_{\text{int}} = \frac{1}{2} \int d^3\mathbf{r} \left(c_0 : \hat{n}^2(\mathbf{r}) : +c_1 : \hat{S}^2(\mathbf{r}) : \right), \qquad (1.46)$$

where $\hat{n}(\mathbf{r}) = \sum_{m=-F}^{F} \hat{\psi}_m^{\dagger}(\mathbf{r})\hat{\psi}_m(\mathbf{r})$ is the particle density operator and $\hat{S}$ is the spin density operator with components

$$\hat{S}_i = \sum_{m,m'=-F}^{F} (\sigma_i)_{mm'} \hat{\psi}_m^{\dagger}(\mathbf{r})\hat{\psi}_{m'}(\mathbf{r}), \qquad (1.47)$$

and σ_i are the spin-1 spin matrices for $i = x, y, z$. The coupling strengths are given by $c_0 = (g_0 + 2g_2)/3$ and $c_1 = (g_2 - g_0)/3$ [51] where $g_{\mathcal{F}_{\text{pair}}}$ ($\mathcal{F}_{\text{pair}} = 0, 2$) are the previously defined interaction strengths in the spin channels.

Similarly, for $F = 2$ the interaction Hamiltonian can be written as

$$\hat{H}_{\text{int}} = \frac{1}{2} \int d^3\mathbf{r} \, c_0 : \hat{n}^2(\mathbf{r}) : +c_1 : \hat{S}^2(\mathbf{r}) : +c_2 \hat{A}_{00}^{\dagger}(\mathbf{r})\hat{A}_{00}(\mathbf{r}), \qquad (1.48)$$

where $\hat{n}(\mathbf{r})$ is defined as previously and $\hat{S}$ is defined as per Eq. (1.47) where σ_i are then taken to be the spin-2 spin matrices for $i = x, y, z$. The coupling strengths are given by $c_0 = (3g_4 + 4g_2)/7$ and $c_1 = (g_4 - g_2)/7$ and $c_2 = (7g_0 - 10g_2 + 3g_4)/7$ (here $\mathcal{F}_{pair} = 0, 2, 4$) [51].

The Single Mode Approximation

In Chaps. 4 and 5, the system under investigation is a small (generally less than 500 atoms) condensate of ^{87}Rb atoms. We thus focus our results from the previous section to this specific application, which allows us to make several assumptions and simplifications. In particular, we focus on the short-time spin-changing dynamics of a condensate initially prepared in the $m_F = 0$ state (we will demonstrate that the particular F state is arbitrary in this instance with respect to the Hamiltonian).

For a ^{87}Rb BEC, the relative values of the different scattering lengths in both hyperfine states allows us to make a number of simplifications to the form of Eqs. (1.46) and (1.48). Firstly, for the $F = 2$ case it has been demonstrated that $c_2 \ll c_0, c_1$ [53] and thus the term proportional to c_2 in Eq. (1.48) can be neglected for short times with respect to the collisional dynamics. Further, for an initial vacuum population in the $m_F = \pm 1, \pm 2$ states the spin-changing processes which gradually populate the $m_F = \pm 2$ states (which first require a sufficient population in the $m_F = \pm 1$ states) will proceed on a much slower time-scale than those which generate atoms in $m_F = \pm 1$ [54]. We can thus effectively ignore all terms involving the $m_F = \pm 2$ field operators and the Hamiltonian is restricted to an effective spin-1 subspace. Under these conditions the Hamiltonians of the $F = 2$ and $F = 1$ hyperfine levels become functionally identical and we thus proceed by considering only Eq. (1.46).

Substituting Eq. (1.46) into Eq. (1.40) and expanding into atomic field operators gives the Hamiltonian in the form [39]

$$\hat{H} = \hat{H}_A + \hat{H}_S + \hat{H}_Z, \tag{1.49}$$

where

$$\hat{H}_S = \int d^3\mathbf{r} \sum_{m=-1}^{1} \hat{\psi}_m^\dagger(\mathbf{r}) \left[-\frac{\hbar^2}{2M} \nabla^2 + V(\mathbf{r}) \right] \hat{\psi}_m(\mathbf{r})$$

$$+ \frac{c_0}{2} \int d^3\mathbf{r} \sum_{m,m'=-1}^{1} \hat{\psi}_m^\dagger(\mathbf{r}) \hat{\psi}_{m'}^\dagger(\mathbf{r}) \hat{\psi}_{m'}(\mathbf{r}) \hat{\psi}_m(\mathbf{r}), \tag{1.50}$$

is known as the symmetric Hamiltonian. This includes the single-particle Hamiltonian as well as elastic (spin-preserving) collisions between the different m_F components. The asymmetric Hamiltonian is given by

$$\hat{H}_A = \frac{c_1}{2} \int d^3\mathbf{r} \left[2\hat{\psi}_0^\dagger(\mathbf{r}) \hat{\psi}_0^\dagger(\mathbf{r}) \hat{\psi}_1(\mathbf{r}) \hat{\psi}_{-1}(\mathbf{r}) + 2\hat{\psi}_1^\dagger(\mathbf{r}) \hat{\psi}_{-1}^\dagger(\mathbf{r}) \hat{\psi}_0(\mathbf{r}) \hat{\psi}_0(\mathbf{r}) \right.$$

$$+ 2\hat{\psi}_0^\dagger(\mathbf{r}) \hat{\psi}_1^\dagger(\mathbf{r}) \hat{\psi}_1(\mathbf{r}) \hat{\psi}_0(\mathbf{r}) + 2\hat{\psi}_0^\dagger(\mathbf{r}) \hat{\psi}_{-1}^\dagger(\mathbf{r}) \hat{\psi}_{-1}(\mathbf{r}) \hat{\psi}_0(\mathbf{r})$$

$$+ \hat{\psi}_1^\dagger(\mathbf{r}) \hat{\psi}_1^\dagger(\mathbf{r}) \hat{\psi}_1(\mathbf{r}) \hat{\psi}_1(\mathbf{r}) + \hat{\psi}_{-1}^\dagger(\mathbf{r}) \hat{\psi}_{-1}^\dagger(\mathbf{r}) \hat{\psi}_{-1}(\mathbf{r}) \hat{\psi}_{-1}(\mathbf{r})$$

$$\left. - 2\hat{\psi}_1^\dagger(\mathbf{r}) \hat{\psi}_{-1}^\dagger(\mathbf{r}) \hat{\psi}_{-1}(\mathbf{r}) \hat{\psi}_1(\mathbf{r}), \right. \tag{1.51}$$

and includes both inelastic spin-changing collisions as well as elastic collisions. Finally, the interaction of the $m_F = \pm 1$ substates with the magnetic field is described by

$$
\hat{H}_Z = \int d^3\mathbf{r} \left\{ \hbar p \left[\hat{\psi}_1^\dagger(\mathbf{r})\hat{\psi}_1(\mathbf{r}) - \hat{\psi}_{-1}^\dagger(\mathbf{r})\hat{\psi}_{-1}(\mathbf{r}) \right] \right.
$$
$$
\left. + \hbar q \left[\hat{\psi}_1^\dagger(\mathbf{r})\hat{\psi}_1(\mathbf{r}) + \hat{\psi}_{-1}^\dagger(\mathbf{r})\hat{\psi}_{-1}(\mathbf{r}) \right] \right\}, \tag{1.52}
$$

where p and q are the forementioned Zeeman coefficients.

In the case that $\hat{H}_S$ is much stronger than $\hat{H}_A + \hat{H}_Z$, requiring $c_0 \gg c_1$, the spatial degrees of freedom will essentially be frozen and dynamics are confined to the internal (spin) degrees of freedom. Assuming $T = 0$ (i.e. $k_B T \ll \hbar\omega$ for a harmonic trapping potential of frequency ω) the field operators for each substate can be expanded in a single-spatial mode $\hat{\psi}_m(\mathbf{r}) \equiv \hat{a}_m \phi_m(\mathbf{r})$, where $\phi_m(\mathbf{r})$ is the ground-state spatial wavefunction (with respect to $\hat{H}_S$) and $\hat{a}_m$ is the annihilation operator of the $m_F = m$ state. This approximation is true for the ^{87}Rb system under consideration as $c_0 \gg c_1$ and the trapping potentials are assumed to be sufficiently tight. The spatial wavefunction $\phi_m(\mathbf{r})$ can be found as the solution of the time-independent mean-field Gross–Pitaevskii equation with respect to $\hat{H}_S$,

$$
\mu\phi_m(\mathbf{r}) = \left[-\frac{\hbar^2}{2M}\nabla^2 + V(\mathbf{r}) + c_0 N|\phi_m(\mathbf{r})|^2 \right] \phi_m(\mathbf{r}), \tag{1.53}
$$

where μ is the chemical potential and N the total number of particles in the condensate. The typical length scale on which the wavefunctions $\phi_m(\mathbf{r})$ of each component differ is known as the spin healing length,

$$
\xi_s = \frac{2\pi\hbar}{\sqrt{2Mc_2\rho}}, \tag{1.54}
$$

where ρ is the density of the trapped condensate. In the case where ξ_s is much larger than the size of the condensate we may make the approximation that $\phi_m(\mathbf{r}) \approx \phi(\mathbf{r})$. This is known as the single mode approximation, wherein all spinor components occupy the same spatial mode. In this case, we have $\hat{\psi}_m(\mathbf{r}) \approx \hat{a}_m \phi(\mathbf{r})$ where $\hat{a}_m$ is the canonical annihilation operator of a particle in the $m = 0, \pm 1$ state and we may integrate out the spatial dimensions of Eqs. (1.51) and (1.52) to give the effective Hamiltonian for the spin dynamics in the single-mode approximation [55],

$$
\hat{H} = \hat{H}_{\text{inel}} + \hat{H}_{\text{el}} + \hat{H}_Z, \tag{1.55}
$$

where

$$\hat{H}_{\text{inel}} = \hbar g \left(\hat{a}_0^\dagger \hat{a}_0^\dagger \hat{a}_1 \hat{a}_{-1} + \hat{a}_1^\dagger \hat{a}_{-1}^\dagger \hat{a}_0 \hat{a}_0 \right), \tag{1.56}$$

$$\hat{H}_{\text{el}} = \hbar g \left(\hat{n}_0 \hat{n}_1 + \hat{n}_0 \hat{n}_{-1} \right), \tag{1.57}$$

$$\hat{H}_Z = \hbar p (\hat{n}_1 - \hat{n}_{-1}) + \hbar q (\hat{n}_1 + \hat{n}_{-1}). \tag{1.58}$$

The coupling constant is defined as $g = (c_2/\hbar) \int d^3\mathbf{r} |\phi(\mathbf{r})|^4$ and $\hat{n}_m$ is the particle number operator for $m = 0, \pm 1$. In our representation we have used that the total atom number $\hat{N} = \hat{a}_0^\dagger \hat{a}_0 + \hat{a}_1^\dagger \hat{a}_1 + \hat{a}_{-1}^\dagger \hat{a}_{-1}$ and number difference $\hat{N}_- = \hat{a}_1^\dagger \hat{a}_1 - \hat{a}_{-1}^\dagger \hat{a}_{-1}$ are conserved quantities.

The inelastic Hamiltonian $\hat{H}_{\text{inel}}$ is responsible for the spin-changing collisions within the spinor BEC, whilst $\hat{H}_{\text{el}}$ and $\hat{H}_Z$ characterize elastic collisions and the Zeeman shift respectively. Combined, these latter two terms act as a population dependent shift in the energy-resonance of the spin-changing collision process. For a spinor BEC initially populating the $m_F = 0$ state, we can further simplify the Hamiltonian into a two-mode model by invoking the undepleted pump approximation for the $m_F = 0$ state, wherein we make the replacement $\hat{a}_0 \to \alpha_0$ where $|\alpha_0|^2 = \langle \hat{a}_0^\dagger \hat{a}_0 \rangle$ is the initial population of the BEC which is assumed to be in a coherent state. For simplicity, we arbitrarily choose α_0 purely real. For the population dynamics this approximation is valid for short times, such that the occupation of the $m_F = \pm 1$ states does not exceed 10 % of the initial $m_F = 0$ population. For an initial vacuum state in the $m_F = \pm 1$ modes we can neglect the linear Zeeman shift as $\hat{n}_1 - \hat{n}_2 = 0$ will be a conserved quantity, and by choosing the quadratic Zeeman shift such that $q = -g|\alpha_0|^2$ and $\hat{H}_Z$ compensates $\hat{H}_{\text{el}}$ (i.e. the spin-changing collisions are on resonance) we find:

$$\hat{H} = \hbar g \alpha_0^2 \left(\hat{a}_1^\dagger \hat{a}_{-1}^\dagger + \hat{a}_1 \hat{a}_{-1} \right), \tag{1.59}$$

which, up to a physically inconsequential phase shift of the pump mode (of $e^{i\pi/4}$), is the same as the squeezing Hamiltonian of Eq. (1.8) where we identify $\kappa_0 \equiv g\alpha_0^2$. Thus, the output state of the spin-changing collision process can be reduced to the two-mode squeezed vacuum in the undepleted pump approximation.

This approximation to the two-mode squeezed vacuum state is pivotal to the work in Chaps. 4 and 5 of this thesis. Specifically, in Chap. 4, motivated by the previous realization of the EPR paradox with massless photons prepared in this state, we investigate the feasibility of a demonstration of the paradox with massive particles by preparing the two-mode squeezed vacuum via spin-changing collisions in a spinor BEC. Furthermore, in Chap. 4 we use the same process to investigate a realization of an atomic SU(1, 1) interferometer, wherein the two-mode squeezed vacuum plays a crucial role in obtaining sensitivity at the Heisenberg limit.

1.5   Background II: Phase-Space Methods

1.5.1   The Wigner Representation

Basic Theory

The earliest and most well-known phase-space distribution is that of the Wigner representation, which was developed by Wigner in 1932 [56]. It is perhaps the most strongly driven by the connection to a classical phase-space and features several, but not all, of the properties of a classical probability distribution.

The basis of the Wigner representation is the Weyl transformation [57],

$$O_W(x, p) = \frac{1}{2\pi} \int dy \, e^{ipx} \langle x - \frac{y}{2} | \hat{O}(x, p) | x + \frac{y}{2} \rangle, \tag{1.60}$$

which maps the quantum operator $\hat{O}(x, p)$ onto the classical phase-space function $O_W(x, p)$ where x and p are canonically conjugate position and momentum variables. Formally, one can then define the relation between the density operator $\hat{\rho}$ of the system and the corresponding Wigner phase-space distribution $W(x, p)$ by the Wigner transformation [57–59]

$$W(x, p) = \frac{1}{2\pi} \int dy \, e^{ipx} \langle x - \frac{y}{2} | \hat{\rho} | x + \frac{y}{2} \rangle, \tag{1.61}$$

which entails a complete representation of the quantum state as a distribution over phase-space variables. Unlike other phase-space representations, the Wigner function has the property that its marginals reflect the true probability distributions of the conjugate variables:

$$P(x) = \int dp \, W(x, p), \tag{1.62}$$

$$P(p) = \int dx \, W(x, p). \tag{1.63}$$

Such a feature is consistent with a true joint-probability distribution for the phase-space variables x and p. However, the interpretation of the Wigner function as a probability distribution has the major flaw that it is not strictly positive for certain quantum states, hence its definition as a quasi-probability distribution. Negativity of the Wigner function is often cited as evidence of non-classicallity of the corresponding quantum state, as the phase-space distribution has no classical analog.

A more general definition of a multimode Wigner function in the coherent state phase-space is given by [60–62]

$$W(\alpha) = \frac{1}{\pi^{2M}} \int d^{2M}\lambda \left[\prod_{n=1}^{M} \exp\left(\lambda_n^* \alpha_n - \lambda_n \alpha_n^*\right) \right] \mathcal{X}_W(\alpha), \tag{1.64}$$

where

$$\mathcal{X}_W(\lambda) = \mathrm{Tr}\left[\hat{\rho} \prod_{n=1}^{M} \exp\left(\lambda_n^* \hat{a}_n - \lambda_n \hat{a}_n^\dagger\right) \right], \tag{1.65}$$

is the characteristic function of the Wigner distribution and $\hat{a}_n$ ($\hat{a}_n^\dagger$) are the annihilation (creation) operators of the $n = 1, 2 \ldots M$ modes. For brevity we adopt the notation $\lambda = (\lambda_1, \ldots, \lambda_M)$ for the integration measure and $\alpha = (\alpha_1, \ldots, \alpha_M)$ for the coherent state amplitudes α_n ($n = 1, 2, \ldots, M$). We interpret $W(\alpha)$ as a quasi-probability distribution for the (complex-valued) amplitudes α_n.

Following the interpretation of $W(\alpha)$ as a quasi-probability distribution, one can calculate expectation values of quantum operators $\hat{O}$ via averages over the phase-space with weighting according to $W(\alpha)$,

$$\langle \hat{O} \rangle_W \equiv \langle O_W \rangle = \int d^{2M}\alpha \; O_W(\alpha) W(\alpha), \tag{1.66}$$

where O_W is the corresponding phase-space representation of $\hat{O}$ obtained by application of the Weyl transformation [see Eq. (1.60)]. The action of the Weyl transformation on operators which can be written as a product of annihilation and creation operators leads to the consequence that averages with respect to the Wigner distribution correspond to quantum expectation values which are symmetrically ordered:

$$\left\langle \prod_{i=1}^{M} \left(\hat{a}_i^\dagger\right)^{m_i} \hat{a}_i^{n_i} \right\rangle_{\mathrm{sym}} = \int d^2\alpha \left[\prod_{i=1}^{M} (\alpha_i^*)^{m_i} \alpha_i^{n_i} \right] W(\alpha). \tag{1.67}$$

The simplest example of this is the calculation of the occupation of a single mode, wherein an average over Wigner phase-space corresponds to

$$\langle \hat{a}_i^\dagger \hat{a}_i \rangle_{\mathrm{sym}} \equiv \frac{1}{2} \langle \hat{a}_i^\dagger \hat{a}_i + \hat{a}_i \hat{a}_i^\dagger \rangle = \int d^{2M}\alpha \; \alpha_i^* \alpha_i W(\alpha). \tag{1.68}$$

Hence, in general one must use the usual bosonic commutation relations $[\hat{a}_i, \hat{a}_j^\dagger] = \delta_{ij}$ to appropriately reorder the Wigner averages and obtain relevant normally-ordered quantum mechanical expectation values, such as those involved in the calculation of second-order correlation functions introduced in the earlier sections of this chapter.

Quantum Evolution Using the Wigner Representation

Beyond its use as an interpretative tool to allow a better understanding and representation of quantum states, the Wigner representation is also a powerful tool to model dynamical problems. In this section we will outline how the evolution of the quantum mechanical density operator in the Schrödinger picture, which is in general intractable, can be mapped to a partial differential equation for the phase-space distribution, which can in general be solved with exact or numerical methods. In particular, we will focus on the application of this representation with respect to the dynamics of spinor condensates, which we will use in detail in Chap. 4.

For a (dissipationless) quantum system subject to unitary evolution, the dynamics of the system are encapsulated by the von Neumann equation for the density operator,

$$i\frac{\partial \hat{\rho}}{\partial t} = \left[\hat{H}, \hat{\rho}\right]. \tag{1.69}$$

This operator equation, in general, is not analytically tractable nor feasible to numerically solve. However, by applying phase-space methods we are able to transform it into an evolution equation for the phase-space distribution, which in general we may solve using numerical techniques.

The basis of the technique is to use a mapping connecting the density matrix and Wigner distribution. In particular, this can be realized by noting that the RHS of the von Neumann equation will be composed of terms like $\hat{a}\hat{\rho}$ and $\hat{\rho}\hat{a}$. Such terms can be generated by application of differential operators on the phase-space distribution (for further detail see, e.g., [61, 62]), to give the relevant operator mappings

$$\hat{a}_i \hat{\rho} \rightarrow \left(\alpha_i + \frac{1}{2}\frac{\partial}{\partial \alpha_i^*}\right) W(\boldsymbol{\alpha}), \tag{1.70}$$

$$\hat{a}_i^\dagger \hat{\rho} \rightarrow \left(\alpha_i^* - \frac{1}{2}\frac{\partial}{\partial \alpha_i}\right) W(\boldsymbol{\alpha}), \tag{1.71}$$

$$\hat{\rho}\hat{a}_i \rightarrow \left(\alpha_i - \frac{1}{2}\frac{\partial}{\partial \alpha_i^*}\right) W(\boldsymbol{\alpha}), \tag{1.72}$$

$$\hat{\rho}\hat{a}_i^\dagger \rightarrow \left(\alpha_i^* + \frac{1}{2}\frac{\partial}{\partial \alpha_i}\right) W(\boldsymbol{\alpha}). \tag{1.73}$$

For terms with multiple creation or annihilation operators, the ordering of the above identities is inside out such that the first differential operator to act on $W(\boldsymbol{\alpha})$ corresponds to the closest operator to $\hat{\rho}$, for example

$$\hat{a}_i^\dagger \hat{a}_i \hat{\rho} \rightarrow \left(\alpha_i^* - \frac{1}{2}\frac{\partial}{\partial \alpha_i}\right)\left(\alpha_i + \frac{1}{2}\frac{\partial}{\partial \alpha_i^*}\right) W(\boldsymbol{\alpha}). \tag{1.74}$$

Substituting these mappings into Eq. (1.69) will convert the master equation into an evolution equation for the phase-space distribution $W(\boldsymbol{\alpha})$. To illustrate this in a concrete way, we specialize our derivation at this point to the case of a spinor condensate in the single-mode approximation (outlined previously in Sect. 1.4.3), which we investigate in depth in Chap. 4. The insights which arise from this example are generically valid for any Hamiltonian which is up to quartic order in creation and annihilation operators. The specific Hamiltonian under consideration is:

$$\hat{H} = \hat{H}_{\text{inel}} + \hat{H}_{\text{el}} + \hat{H}_Z, \tag{1.75}$$

where

$$\hat{H}_{\text{inel}} = \hbar g \left(\hat{a}_0^\dagger \hat{a}_0^\dagger \hat{a}_1 \hat{a}_{-1} + \hat{a}_1^\dagger \hat{a}_{-1}^\dagger \hat{a}_0 \hat{a}_0 \right), \tag{1.76}$$

$$\hat{H}_{\text{el}} = \hbar g \left(\hat{n}_0 \hat{n}_1 + \hat{n}_0 \hat{n}_{-1} \right), \tag{1.77}$$

$$\hat{H}_Z = \hbar q (\hat{n}_1 + \hat{n}_{-1}). \tag{1.78}$$

The coupling constant g is characterised by s-wave scattering of the atoms, whilst q parametrizes the quadratic Zeeman shift (see Sect. 1.4.3 for more detail). Substitution of Eq. (1.75) into the von Neumman equation [Eq. (1.69)] and application of the operator mappings gives the result

$$\frac{dW(\alpha)}{dt} = \left\{ \sum_{n=0}^{2} \left(\frac{\partial}{\partial \alpha_n} A_n + \frac{\partial}{\partial \alpha_n^*} A_n^* \right) + \mathcal{O}\left(\frac{\partial^3}{\partial \alpha_n^3} \right) \right\} W(\alpha), \tag{1.79}$$

where $\mathbf{A}$ is the drift vector, characterised in this case by

$$A_0 = -ig \left[2\alpha_1 \alpha_2 \alpha_0^* + \left(|\alpha_1|^2 + |\alpha_2|^2 \right) \alpha_0 \right], \tag{1.80}$$

$$A_1 = -ig \left[\alpha_0^2 \alpha_2^* + \left(|\alpha_0|^2 + q/g \right) \alpha_1 \right], \tag{1.81}$$

$$A_2 = -ig \left[\alpha_0^2 \alpha_1^* + \left(|\alpha_0|^2 + q/g \right) \alpha_2 \right]. \tag{1.82}$$

Note that in this case α_2 corresponds to the mode $\hat{a}_{-1}$ and the phase-space variables span $\alpha \equiv (\alpha_0, \alpha_1, \alpha_2)$. More generally, for a system with M modes $\mathbf{A}$ will be a vector of length M. The third-order derivative terms in Eq. (1.79) arise due to the quartic products of annihilation and creation operators in $\hat{H}_{\text{inel}}$ and $\hat{H}_{\text{el}}$. In general, including the case here, such terms render Eq. (1.79) intractable. To overcome this, one invokes the 'truncated Wigner approximation' (TWA), wherein differential terms of third-order and higher are explicitly neglected. This assumption is valid in general, for systems where the average occupation is large $N \gg 1$, or for multi-mode systems where the total number of particles greatly exceeds the number of modes (M) $N \gg M$ [61, 62]. In this case, one generally uses a broad argument that the contribution of the third-order terms is negligible, at least for short times, as they scale as $1/N^{3/2}$. Theoretical study has examined the applicability and robustness of this criteria [63], however, inevitably the truncation error is difficult to quantify without another solution method with which to benchmark.

The truncated Wigner approximation thus renders Eq. (1.79) into the form of a classical Liouville equation [61, 62],

$$\frac{dW(\alpha)}{dt} = \left\{ \sum_{n=0}^{2} \left(\frac{\partial}{\partial \alpha_n} A_n + \frac{\partial}{\partial \alpha_n^*} A_n^* \right) \right\} W(\alpha), \tag{1.83}$$

which can be solved by the method of characteristics. This method reduces the Liouville equation for the distribution function to an equivalent deterministic differential equation for the phase-space variables,

$$\frac{d\alpha_n}{dt} = A_n \quad (n = 0, 1, 2), \tag{1.84}$$

with stochastic initial conditions corresponding to an appropriately sampled Wigner distribution of the initial state. Under the forementioned approximations of a dissipationless system and truncation of the higher-order terms in the evolution equation for $W(\alpha)$, the differential equation for α_n (α_n^*) always correspond to the Heisenberg equation of motion for $\hat{a}_n$ ($\hat{a}_n^\dagger$) with replacement of the operators by their c-number counterparts (coherent amplitudes), as can easily be seen from inspection of Eqs. (1.80)–(1.82).

As the differential equation is deterministic in this approximation, the 'quantumness' of the problem is encapsulated by the stochastic initial conditions of α. These are chosen to reflect the underlying Wigner distribution of the initial state, which in the TWA must be strictly positive so that it can be mapped to a classical probability distribution which can then be sampled [64]. For the system in question the initial Wigner distribution is assumed to be seperable such that $W(\alpha) = W_0(\alpha_0)W_1(\alpha_1)W_2(\alpha_2)$ where $W_n(\alpha_n)$ is the single-mode Wigner function for modes $n = 0, 1, 2$ and hence the stochastic initial condition of each mode can be sampled independently.

We assume that the spinor condensate is initially prepared purely in the $\hat{a}_0$ mode and can be modelled well by a coherent state of amplitude β_0, such that $W_0(\alpha_0) = \frac{2}{\pi}\exp(-2|\alpha_0 - \beta_0|^2)$. Following the prescription of Ref. [64] this distribution can be sampled by choosing the initial condition $\alpha_0(0) = \beta_0 + \eta_0/\sqrt{2}$ where η_0 is a complex source of white noise such that $\langle\eta_0\rangle = 0$ and $\langle\eta_0^2\rangle = 1$.

In the investigation of Chap. 4, we consider the $\hat{a}_{\pm 1}$ modes to be in one of three initial conditions: (i) vacuum state, (ii) thermal state or (iii) coherent state. The third of these is identical to that outlined above for the $\hat{a}_0$ mode, albeit for a different coherent amplitude, and so we will not repeat it here. Similarly, the vacuum state can be trivially modelled as a coherent state with nil amplitude, such that an appropriate initial condition for the phase-space variables is $\alpha_n(0) = \eta_n/\sqrt{2}$ where $n = 1, 2$ and η_n ($n = 1, 2$) are complex independent sources of white noise with $\langle\eta_n\rangle = 0$ and $\langle\eta_n\eta_m\rangle = \delta_{mn}$. Finally, the Wigner distribution for a thermal state is given by [64]

$$W_n(\alpha_n) = \frac{1}{\pi}\frac{1}{\bar{n} + 1/2}e^{-|\alpha_n|^2/(\bar{n}+1/2)} \tag{1.85}$$

where $\bar{n} = \langle\hat{a}_n^\dagger\hat{a}_n\rangle$ is the average occupation and we consider $n = 1, 2$ here. Such a distribution can be sampled appropriately by choosing the initial condition $\alpha_n(0) = \sqrt{\bar{n} + 1/2}\eta_n e^{2\pi\xi_n}$ for $n = 1, 2$ where η_n is a complex source of white noise as before and ξ_n is a random real variable uniformly distributed in the interval $[0, 1]$.

We note that for a full multi-mode treatment of a spinor BEC taking into account the spatial degrees of freedom, the initial state of the system would need to be fully characterised by a Bogoliubov analysis of the low-lying excitations around the condensate [65, 66]. Furthermore, for a spinor system this Bogoliubov analysis would also need to take into account the spinor excitations (spin waves) to fully characterise the quantum noise [67] of the initial state. However, in this thesis we consider spinor systems which are very well described in the single-mode approximation (see Sect. 1.4.3), and specifically the spin-healing length is negligible when compared to the characteristic length scale of the spatial wavefunction. Under these conditions, the TWA treatment of the quantum dynamics of the simple few-mode model encapsulates the relevant physics on which we focus.

With appropriate choice of initial quantum noise, the evolution of the ensemble of stochastically-seeded trajectories can then be interpreted as the evolution of the total phase-space distribution in time. Hence, the final Wigner function can be easily reconstructed by considering the distribution of the final ensemble of solutions $\alpha(t)$, specifically by binning the values (with sufficient resolution) in the $2M$-dimensional complex space. This procedure can always be carried out as the truncated Wigner approximation intrinsically guarantees that the non-negativity of the initial Wigner function is preserved throughout the evolution.

The deterministic evolution combined with stochastic initial conditions tempts an interpretation in terms of classical phase-space distributions. However, there are several important differences that render this less useful. In particular, classically the uncertainty in initial conditions is due to imperfect knowledge of the underlying classical state and thus each individual trajectory [solution of $\alpha(t)$] retains meaning. In contrast, individual stochastic trajectories or equivalently points in the quantum phase-space with simultaneously well defined $\mathrm{Re}(\alpha_n)$ and $\mathrm{Im}(\alpha_n)$ for $n = 1, 2 \ldots M$ cannot physically exist, as $\mathrm{Re}(\alpha_n)$ and $\mathrm{Im}(\alpha_n)$ are related to *incompatible* observables $(\hat{a}_n + \hat{a}_n^\dagger)/2$ and $i(\hat{a}_n^\dagger - \hat{a}_n)/2$ respectively. As a consequence of the Heisenberg uncertainty principle for these observables, simultaneous values of $\mathrm{Re}(\alpha_n)$ and $\mathrm{Im}(\alpha_n)$ are inconsistent with the standard interpretation of quantum mechanics. In a similar vein, interpreting the finite width of the Wigner distribution as due to imperfect knowledge of the underlying state would be akin to a hidden-variable interpretation, whereas this is actually a reflection of the uncertainty in the canonically conjugate variables. We discuss this common interpretation of the Wigner function in more detail in Chap. 6.

Symmetrically-ordered expectation values $\langle \prod_{i=0}^{2} (\hat{a}_i^\dagger)^{n_i} \hat{a}_i^{m_i} \rangle_{\mathrm{sym}}$ are trivially calculated according to Eq. (1.67), which corresponds to averaging $\prod_{i=0}^{2} (\alpha_i^*)^{n_i} \alpha_i^{m_i}$ over the ensemble of trajectories. Neglecting truncation error, the error in expectation values will be only due to the finite size of the ensemble and the sampling error will scale as $1/N$ where N is the number of trajectories.

For the process of spin-changing collisions we are specifically interested in normally-ordered correlation functions such as $\langle \hat{n}_i \rangle$ and $G_{ij}^{(2)} = \langle : \hat{n}_i \hat{n}_j : \rangle$ ($i, j = 0, \pm 1$ where $i, j = -1$ corresponds to the field α_2). The population of each mode can trivially be calculated according to

$$\langle \hat{a}_i^\dagger \hat{a}_i \rangle \equiv \langle |\alpha_i|^2 \rangle_{\mathrm{stoch}} - \frac{1}{2}, \tag{1.86}$$

where the subtraction of $1/2$ is a correction due to the symmetric ordering and $\langle \cdots \rangle_{\text{stoch}}$ refers to an average over phase-space trajectories. Similarly, the relevant cross- and auto-correlations can be calculated respectively as:

$$G_{ij}^{(2)} \equiv \langle |\alpha_i|^2 |\alpha_j|^2 \rangle_{\text{stoch}} - \frac{1}{2} \left(\langle |\alpha_i|^2 \rangle_{\text{stoch}} + \langle |\alpha_j|^2 \rangle_{\text{stoch}} \right) + \frac{1}{4} \quad (\text{for } i \neq j), \quad (1.87)$$

$$G_{ii}^{(2)} \equiv \langle |\alpha_i|^4 \rangle_{\text{stoch}} - 2 \langle |\alpha_i|^2 \rangle_{\text{stoch}} + \frac{1}{2}. \quad (1.88)$$

In Fig. 1.5 we plot results of an example calculation using the truncated Wigner approximation for this system. We prepare an initial coherent state in the $n = 0$ mode, such that the initial population is $\langle \hat{a}_0^\dagger \hat{a}_0 \rangle = |\beta_0|^2 = 50$, whilst the sidemodes are prepared in an initial vacuum state. The coupling coefficient is chosen to be $g = 0.05$ and the quadratic Zeeman shift is $q = -gN_0$. We calculate the mean population of the sidemodes $\langle \hat{n}_1 + \hat{n}_{-1} \rangle$ and the normalized second-order correlations $g_{ij}^{(2)} = G_{ij}^{(2)} / (\langle \hat{n}_i \rangle \langle \hat{n}_j \rangle)$ using the TWA and compare these results to calculations using the exact diagonalization method in the Fock state basis. We find excellent agreement for short times until a small discrepancy in $\langle \hat{n}_1 + \hat{n}_{-1} \rangle$ at the first maximum. Breakdown of the TWA method is more clearly seen after $t \sim 0.3$, wherein there is a clear difference between the exact results and the TWA calculation. This deviation

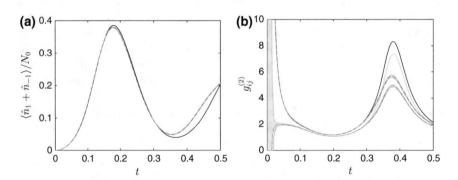

Fig. 1.5 **a** Mean occupation of the sidemodes $\langle \hat{n}_1 + \hat{n}_{-1} \rangle$ calculated using the truncated Wigner approximation [Eqs. (1.84), *dashed red line*] compared to that calculated using exact diagonalization of the Hamiltonian (*solid black line*). **b** Second-order correlation functions calculated in the truncated Wigner approximation: $g_{1,-1}^{(2)}$ (*dashed red line*) and $g_{1,1}^{(2)} = g_{-1,-1}^{(2)}$ (*dot-dashed magenta line*). These are compared to results calculated using exact diagonalization (*solid black and dotted blue lines* respectively). Sampling error (two standard deviations) is indicated by the shading for the TWA results. Parameters of the calculation are discussed in the main text. In both **a** and **b** good agreement is found for short times. Early differences in the $g_{ij}^{(2)}$ results are due to large sampling error due to the low population of the sidemodes, which means the numerator and denominator of $g_{ij}^{(2)}$ are both dominated by stochastic noise. Discrepancies occur near $t \sim 0.15$ for calculations of the mean population $\langle \hat{n}_1 + \hat{n}_2 \rangle$, whilst a clear difference between the methods is seen in both the mean population and second-order correlations for $t \gtrsim 0.3$. The onset of these discrepancies is due to the truncation error introduced by neglecting the third-order derivatives in the exact evolution equation for $W(\alpha)$ [Eq. (1.79)]

is entirely attributable to truncation error, which is introduced upon discarding third-order derivatives and higher from the exact evolution equation for $W(\alpha)$ [Eq. (1.79)].

Beyond calculation of correlation functions, it is also possible to evaluate the particle number distribution of the system via the overlap formula [59],

$$P_{n_1,n_2...n_M} = \int d^{2M}\alpha \, W(\alpha) \prod_{i=1}^{M} W_{|n_i\rangle}(\alpha_i), \tag{1.89}$$

where

$$W_{|n_i\rangle}(\alpha_i) = \frac{2}{\pi}(-1)^{n_i} e^{-2|\alpha_i|^2} L_{n_i}(4|\alpha_i|^2), \tag{1.90}$$

is the Wigner function of the Fock state $|n_i\rangle$ and $L_{n_i}(x)$ is the n_i-th order Laguerre polynomial. It is easy to see that Eq. (1.89) is operationally equivalent to evaluating the average of $\langle \prod_{i=1}^{M} W_{|n_i\rangle}(\alpha_i) \rangle_{\text{stoch}}$ over the ensemble of trajectories. Such a computation is in general non-trivial for highly occupied states or those with a sufficiently broad number distribution as it requires evaluation of high-order Laguerre polynomials with large arguments. Usually, computational techniques such as quadruple precision will be required to overcome numerical issues for $n_i \gtrsim 256$.

In Chap. 6 we outline another operational method which can be implemented to calculate an approximation to $P_{n_1,n_2...n_M}$ under certain conditions. Specifically, focusing on the single-mode case P_n, we outline how by assuming that individual stochastic trajectories can be regarded as faithful representations of experimental trials—a point which is touched upon earlier in this subsection—one is motivated to bin the occupation of individual stochastic trajectories $n_j \equiv |\alpha_j|^2 - 1/2$, where the subscript j denotes the trajectory index, into a probability distribution $\tilde{P}_n$ which can then be compared to P_n. We demonstrate in Chap. 6 that there is indeed a close quantitative correspondence between P_n and $\tilde{P}_n$ for a wide range of states, which indicates that $\tilde{P}_n$ is a simple computational alternative in practice.

1.5.2 The Positive-P Representation

Basic Theory

Another class of phase-space distributions are given by the P-representations. In contrast to the Wigner function, which has the drawback of being negative in certain cases, the P functions have the favourable property that the distribution function is strictly positive. In general, the familiy of P representations are defined as an expansion of the density matrix over the off-diagonal coherent state basis [61, 62, 68]

$$\hat{\rho} = \int \hat{\Lambda}(\alpha, \beta) P(\alpha, \beta) d\mu(\alpha, \beta), \tag{1.91}$$

where $d\mu(\alpha, \beta)$ is the integration measure which defines the particular P-representation and we define

$$\hat{\Lambda}(\alpha, \beta) = \frac{|\alpha\rangle\langle\beta^*|}{\langle\beta^*|\alpha\rangle} \tag{1.92}$$

as the off-diagonal coherent state projection operator. The coherent amplitudes which span the phase-space are defined via the vector notation $\alpha = (\alpha_1, \alpha_2, \ldots, \alpha_M)$ and $\beta = (\beta_1, \beta_2, \ldots, \beta_M)$ where M is the number of modes in the system.

In the simplest case, where $d\mu(\alpha, \beta) = d^{2M}\alpha d^{2M}\beta\delta^{(M)}(\alpha - \beta)$ one recovers the well-known (diagonal) Glauber–Sudershan P-representation. However, the non-negativity of this distribution comes at the cost that it is not necessarily well defined for all density matrices. In particular, for some states the distribution function is more singular than a δ-function [68].

A solution to this problem was found by Drummond and Gardiner [68], where at the cost of effectively doubling the phase-space with $d\mu(\alpha, \beta) = d^{2M}\alpha d^{2M}\beta$, they were able to define a strictly non-negative distribution function which is well defined for any state, known as the positive-P representation. In contrast to the Wigner representation where one makes the effective correspondence $\hat{a}_i \leftrightarrow \alpha_i$ and $\hat{a}_i^\dagger \leftrightarrow \alpha_i^*$ for $i = 1 \ldots M$, in the doubled phase-space of the positive-P representation we associate $\hat{a}_i \leftrightarrow \alpha_i$ and $\hat{a}_i^\dagger \leftrightarrow \beta_i$ where we emphasize that α_i and β_i are *independent* variables which are *not* related by complex conjugation.

Similar to the Wigner distribution, expectation values of quantum operators may be obtained by appropriate averaging of the phase-space variables with respect to the weighting of the $P(\alpha, \beta)$ function. In the case of the positive-P representation, these averages correspond to normally-ordered products of annihilation and creation operators

$$\left\langle : \prod_{i=1}^{M}(\hat{a}_i^\dagger)^{n_i}\hat{a}_i^{m_i} : \right\rangle = \int d^2\alpha d^2\beta \left[\prod_{i=1}^{M}\beta_i^{m_i}\alpha_i^{n_i}\right] P(\alpha, \beta), \tag{1.93}$$

where the colon notation indicates normal ordering (i.e. all creation operators are to the left of all annihilation operators).

Quantum Evolution Using the Positive-P Representation

Having appropriately defined the positive-P distribution we will show in this section how the phase-space representation can be used to solve the dynamics of the quantum mechanical problems. Identically to the Wigner representation, the basis of the method is to map the intractable master equation for the density operator to a partial differential equation for the evolution of $P(\alpha, \beta)$ – a Fokker-Planck equation – which can be mapped to a set of stochastic differential equations and solved numerically. In particular, we will focus on the application of this method to the topic of spontaneous four-wave mixing in colliding condensates, which forms the basis of Chaps. 2 and 3.

The evolution of the density operator can be rewritten in terms of the positive-P representation [61, 62, 69],

$$\frac{\partial \hat{\rho}}{\partial t} = \int \hat{\Lambda}(\alpha, \beta) \frac{\partial P(\alpha, \beta)}{\partial t} d^2\alpha d^2\beta, \tag{1.94}$$

and thus using the (dissipationless) von Neumann equation for the density operator [Eq. (1.69)] we have [61, 62, 69],

$$\int \hat{\Lambda}(\alpha, \beta) \frac{\partial P(\alpha, \beta)}{\partial t} d^2\alpha d^2\beta = \int \left[\hat{H}, \hat{\Lambda}(\alpha, \beta) \right] P(\alpha, \beta) d^2\alpha d^2\beta, \tag{1.95}$$

Similar to the derivation for the Wigner representation of Sect. 1.5.1, the relevant mappings between the creation and annihilation operators and the phase-space variables for the positive-P representations can be written in terms of the projection operator [61, 62, 70]:

$$\hat{a}_i \hat{\Lambda}(\alpha, \beta) \to \alpha_i \hat{\Lambda}(\alpha, \beta), \tag{1.96}$$

$$\hat{a}_i^\dagger \hat{\Lambda}(\alpha, \beta) \to \left(\beta_i + \frac{\partial}{\partial \alpha_i} \right) \hat{\Lambda}(\alpha, \beta), \tag{1.97}$$

$$\hat{\Lambda}(\alpha, \beta) \hat{a}_i \to \left(\alpha_i - \frac{\partial}{\partial \beta_i} \right) \hat{\Lambda}(\alpha, \beta), \tag{1.98}$$

$$\hat{\Lambda}(\alpha, \beta) \hat{a}_i^\dagger \to \beta_i \hat{\Lambda}(\alpha, \beta). \tag{1.99}$$

Substitution of these mappings into Eq. (1.95) and integration by parts to shift the differential operators onto $P(\alpha, \beta)$ will lead to a generalized Fokker-Planck equation for the phase-space distribution. Restricting our consideration to Hamiltonians which are no higher than quartic in creation and annihilation operators, the form of the equation will be an actual Fokker-Planck equation, with no higher than second-order derivative terms,

$$\frac{d}{dt} P(\alpha) = \left[\sum_i \frac{\partial}{\partial \alpha_i} A_i + \frac{1}{2} \sum_{i,j} \frac{\partial^2}{\partial \alpha_i \partial \alpha_j} D_{ij} \right] P(\alpha), \tag{1.100}$$

where $\mathbf{A}$ is the drift vector of length $2M$, $\mathbf{D}$ the diffusion matrix of dimensions $2M \times 2M$ and we have used the vector notation $\alpha \equiv (\alpha_1, \alpha_2, \ldots, \alpha_M, \beta_1, \ldots, \beta_M)$ for simplicity. Due to the nature of the positive-P representation the diffusion matrix $\mathbf{D}$ is guaranteed to be positive-definite [61, 70], implying Eq. (1.100) is always a valid Fokker-Planck equation. We make the important observation that for this derivation to be valid, we assume that when integrating by parts the boundary terms of the distribution $P(\alpha, \beta)$ vanish (in particular this requires that the tails of the phase-space distribution decay exponentially). This assumption plays an important role in the practical application of the distribution to quantum mechanical problems and we will discuss consequences of this assumption in further detail later in this section.

The Fokker-Planck equation for $P(\alpha)$ [Eq. (1.100)] can be solved by mapping the equation to a set of Îto stochastic differential equations, such that:

$$\frac{d\alpha_j}{dt} = -A_j + \sum_k B_{jk}\xi_k(t). \tag{1.101}$$

Here $\mathbf{B}$ is a $2M \times \mathcal{N}$ matrix which is defined as the (non-unique) decomposition of the diffusion matrix $\mathbf{D} = \mathbf{BB}^T$ and $\boldsymbol{\xi}(t)$ is a vector of $\mathcal{N}$ independent, real Gaussian white noise terms with $\langle \xi_j(t) \rangle = 0$ and $\langle \xi_j(t)\xi_k(t') \rangle = \delta_{jk}\delta(t - t')$ where $j, k = 1, 2 \ldots \mathcal{N}$. These Îto equations may also be converted into Stratonovich form [61, 62], to which the usual rules of differential calculus apply, simplifying their integration with standard numerical methods (see, e.g., Ref. [71]).

For illustration, we show how this method can be implemented to the problem of colliding condensates. In this thesis we use the positive-P formalism to calculate correlation functions for a system of $\sim 10^5$ particles on a spatial lattice of $\sim 10^7$ modes. Such a problem is clearly beyond exact techniques due to the enormous Hilbert space, however, the positive-P representation allows us to map the intractable master equation to a Fokker-Planck equation which can be solved using stochastic techniques.

The Hamiltonian describing the process is:

$$\hat{H} = \int d^3\mathbf{r} \left\{ \hat{\psi}^\dagger(\mathbf{r}) \left[-\frac{\hbar^2}{2m}\nabla^2 \right] \hat{\psi}(\mathbf{r}) + \frac{U}{2}\hat{\psi}^\dagger(\mathbf{r})\hat{\psi}^\dagger(\mathbf{r})\hat{\psi}(\mathbf{r})\hat{\psi}(\mathbf{r}) \right\}, \tag{1.102}$$

where $U = 4\pi\hbar^2 a/m$ is the interaction strength characterized by the s-wave scattering length a between atoms (of mass m) described by the bosonic creation (annihilation) field operators $\hat{\psi}^\dagger$ ($\hat{\psi}$). In order for the assumed delta-function form of the interaction potential $U(\mathbf{r} - \mathbf{r}') = U\delta^{(3)}(\mathbf{r} - \mathbf{r}')$ to be valid, one must impose a finite momentum cutoff $\mathbf{k}_{max}$ such that $|\mathbf{k}_{max}| \gg 1/a$ [72]. For discrete lattice models, which we will discuss below, this can be physically ensured by choosing an appropriate lattice spacing in position space when the continuous system is discretized. Enforcing this momentum cutoff then ensures the interaction strength U is given by the formula above and does not require explicit renormalization to avoid UV divergences [72].

To solve this problem in the positive-P representation we are required to discretize the continuous Hamiltonian of Eq. (1.102) onto a three-dimensional spatial lattice, such that appropriate creation and annihilation operators can be defined, which can then be mapped to the phase-space variables. We characterise the lattice points by the vector $\boldsymbol{i} = (i_x, i_y, i_z)$ where $i_x = 1, 2 \ldots N_x$ and similarly for i_y and i_z, with spatial co-ordinates $\mathbf{r}_i = (i_x\Delta x, i_y\Delta y, i_z\Delta z)$ where Δx is the lattice spacing in the x direction and similarly for y and z. For simplicity we choose the lattice to have $N_x = N_y = N_z = N$ points along each dimension of total lengths $L_x = N\Delta x$, $L_y = N\Delta y$ and $L_z = N\Delta z$, such that there are N^3 lattice points (modes) in total (for completeness we note this is not the case in Chaps. 2 and 3, however, the following results are trivially generalized). Subsequently, we can define new bosonic annihilation operators on the lattice $\hat{a}_i \equiv (\Delta x \Delta y \Delta z)^{1/2} \hat{\psi}(\mathbf{r}_i)$ such that $[\hat{a}_i, \hat{a}_j^\dagger] = \delta_{ij}$. The discretized Hamiltonian can then be written as [69, 73]

$$\hat{\mathcal{H}} = \sum_{i,j} \hbar \omega_{ij} \hat{a}_i^\dagger \hat{a}_j + \hbar \mathcal{U} \sum_i \hat{a}_i^\dagger \hat{a}_i^\dagger \hat{a}_i \hat{a}_i, \qquad (1.103)$$

where

$$\mathcal{U} = \frac{U}{2\hbar} \Delta x \Delta y \Delta z, \qquad (1.104)$$

is the discretized coupling constant and [69]

$$\omega_{ij} = \frac{\hbar}{2mN^3} \sum_{\mathbf{k}} |\mathbf{k}|^2 e^{i\mathbf{k} \cdot (\mathbf{r}_i - \mathbf{r}_j)}, \qquad (1.105)$$

is a linear coupling between sites. This last term arises due to the discretization of the spatial derivative in the kinetic energy term via a Fourier transform to momentum space. The summation over plane-wave momentum modes is defined such that $\mathbf{k} = (k_x, k_y, k_z)$ where for N odd $k_i = (-k_i^{\max}, -k_i^{\max} + \Delta k_i, \ldots, k_i^{\max})$ for $\Delta k_i = 2\pi/L_i$ and $k_i^{\max} = (N-1)\Delta k_i/2$ $(i = x, y, z)$ for periodic boundary conditions. Similarly, for N even we have $k_i = (-k_i^{\max}, -k_i^{\max} + \Delta k_i, \ldots, k_i^{\max} - \Delta k_i)$ $(i = x, y, z)$. As discussed previously, for the implicitly assumed δ-function form of the interaction potential to be valid we require that $k_i^{\max}$ is chosen such that $a \gg 1/k_i^{\max}$. In terms of the lattice spacing in position space, this requirement is equivalent to choosing N such that $\Delta x, \Delta y, \Delta z \gg a$.

Evaluation of the von Neumann equation [Eq. (1.95)] with respect to the discretized Hamiltonian $\hat{\mathcal{H}}$ [Eq. (1.103)], substitution of the mappings of Eqs. (1.96)–(1.99) into the ensuing master equations, and integration by parts leads to a Fokker-Planck equation of the form of Eq. (1.100). This equation is solved by mapping to a set of coupled Îto stochastic differential equations, which in this instance have the form:

$$\frac{d\alpha_j}{dt} = -i \left[\sum_l (\omega_{jl} \alpha_l) + 2\mathcal{U} \alpha_j^2 \beta_j + \sqrt{2i\mathcal{U}} \alpha_j \xi_{1,j}(t) \right], \qquad (1.106)$$

$$\frac{d\beta_j}{dt} = i \left[\sum_l (\omega_{jl} \beta_l) + 2\mathcal{U} \beta_j^2 \alpha_j + \sqrt{2i\mathcal{U}} \beta_j \xi_{2,j}(t) \right], \qquad (1.107)$$

where $\xi_{k,j}(t)$ $(k = 1, 2$ and $j = 1, 2 \ldots M)$ are independent sources of real Gaussian noise with $\langle \xi_{kj}(t) \rangle = 0$ and $\langle \xi_{kj}(t) \xi_{ml}(t') \rangle = \delta_{km} \delta_{jl} \delta(t - t')$.

We can define less unwieldy stochastic fields by the relation $\Psi(\mathbf{r}) \equiv \Delta x \Delta y \Delta z \alpha_{\mathbf{r}_i}$ and $\tilde{\Psi}(\mathbf{r}) \equiv \Delta x \Delta y \Delta z \beta_{\mathbf{r}_i}$. Then, by taking $\Delta x \to 0$ and similarly for Δy and Δz we can recover stochastic equations for the continuous fields $\Psi(\mathbf{r}, t)$ and $\tilde{\Psi}(\mathbf{r}, t)$,

$$\frac{d\Psi(\mathbf{r}, t)}{dt} = i\frac{\hbar}{2m}\nabla^2\Psi(\mathbf{r}, t) - i\frac{U}{\hbar}\Psi(\mathbf{r}, t)^2\tilde{\Psi}(\mathbf{r}, t) + \sqrt{-i\frac{U}{\hbar}\Psi(\mathbf{r}, t)}\xi_1(\mathbf{r}, t), \quad (1.108)$$

$$\frac{d\tilde{\Psi}(\mathbf{r}, t)}{dt} = -i\frac{\hbar}{2m}\nabla^2\tilde{\Psi}(\mathbf{r}, t) + i\frac{U}{\hbar}\tilde{\Psi}(\mathbf{r}, t)^2\Psi(\mathbf{r}, t) + \sqrt{i\frac{U}{\hbar}\tilde{\Psi}(\mathbf{r}, t)}\xi_2(\mathbf{r}, t), \quad (1.109)$$

where $\xi_j(\mathbf{r}, t)$ ($j = 1, 2$) is a source of real Gaussian noise with $\langle\xi_j(\mathbf{r}, t)\rangle = 0$ and $\langle\xi_j(\mathbf{r}, t)\xi_k(\mathbf{r}', t')\rangle = \delta_{jk}\delta^{(3)}(\mathbf{r} - \mathbf{r}')\delta(t - t')$.

The Stratonovich form of these stochastic equations (which are equivalent to the Îto form in this example up to a physically irrelevant global phase) may be easily simulated on a computer via standard numerical techniques. Similar to the Wigner representation, the initial condition for the fields $\Psi(\mathbf{r}, 0)$ and $\tilde{\Psi}(\mathbf{r}, 0)$ is stochastically sampled by mapping the underlying P distribution to a classical probability distribution. For the example of spontaneous four-wave mixing of matter waves, we treat the initial unsplit condensate as a coherent state, which in the P representation corresponds to a delta function. Consequently, in contrast to the Wigner function example, we find that for this case there is no stochastic noise in the initial condition and we have $\Psi(\mathbf{r}, 0) = \tilde{\Psi}(\mathbf{r}, 0) = \sqrt{\rho(\mathbf{r}, 0)/2}e^{i\mathbf{k}_0 \cdot \mathbf{r}} + \sqrt{\rho(\mathbf{r}, 0)/2}e^{-i\mathbf{k}_0 \cdot \mathbf{r}}$ where $\rho(\mathbf{r}, 0)$ is the density profile of the initial (unsplit) condensate (whose coherent amplitude we take to be real without loss of generality) and $\pm\mathbf{k}_0$ are the momenta of the two counter-propagating halves of the split condensate.

In the positive-P representation, normally-ordered quantum mechanical expectation values of $\hat{\psi}$ and $\hat{\psi}^\dagger$ can be calculated by stochastic averages of the corresponding fields ψ and $\tilde{\psi}$ respectively,

$$\left\langle\left(\hat{\psi}^\dagger\right)^n\hat{\psi}^m\right\rangle = \langle\tilde{\Psi}^n\Psi^m\rangle_{\text{stoch}}. \quad (1.110)$$

Whilst it may appear at this point that the positive-P representation has overcome the problems suffered by the Wigner representation, namely truncation error and negativity of the initial distribution, there are different issues which we must address for the positive-P representation. Specifically, when the von Neumman equation for the density operator is mapped to a Fokker-Planck equation for the P distribution we assume that the boundary terms (which arise during the integration by parts) vanish and then the mapping is exact. For systems which are only quadratic in creation and annihilation operators, this condition is trivially satisfied as the boundary terms do not exist [70, 74]. However, for more general systems non-negligible boundary terms may in practice arise after some simulation time. Fortunately, the existence of such terms is usually accompanied by easily identifiable numerical signatures such as a sudden increase in sampling errror (due to divergent trajectories) and spiking of trajectories which are both in general a consequence of the development of power-law tails in the probability distribution [75].

In the case of colliding condensates, an empirical estimate of the time at which boundary terms may appear is given by [25, 74]

$$t_{\text{sim}} = \frac{5m\Delta x\Delta y\Delta z}{8\pi\hbar a[\rho(0, 0)]^{2/3}} \quad (1.111)$$

which for the parameters investigated in Chaps. 2 and 3 is of the same order of magnitude as the collision duration (typically $\sim$50–100 μs).

To overcome this issue and simulate the collision for the full duration one can consider implementing a linearization of the positive-P equations around some appropriate mean-field solution. Specifically, we consider $\Psi(\mathbf{r}, t) \equiv \psi(\mathbf{r}, t) + \delta(\mathbf{r}, t)$ and $\tilde{\Psi}(\mathbf{r}, t) \equiv \psi(\mathbf{r}, t)^* + \tilde{\delta}(\mathbf{r}, t)$ where $\delta(\mathbf{r}, t)$ and $\tilde{\delta}(\mathbf{r}, t)$ are the lowest order perturbations to the mean-field solution $\psi(\mathbf{r}, t) \equiv \langle \hat{\psi}(\mathbf{r}, t) \rangle = \langle \Psi(\mathbf{r}, t) \rangle_{\text{stoch}}$ $[\psi(\mathbf{r}, t)^* \equiv \langle \hat{\psi}^\dagger(\mathbf{r}, t) \rangle = \langle \tilde{\Psi}(\mathbf{r}, t) \rangle_{\text{stoch}}]$ of Eqs. (1.108) and (1.109). In particular, $\psi(\mathbf{r}, t)$ describes the split source condensates which obey the deterministic Gross–Pitaevskii equation and are treated in the undepleted pump approximation (see Sect. 1.4.2 for further details). The incoherent stochastic fields $\delta(\mathbf{r}, t)$ and $\tilde{\delta}(\mathbf{r}, t)$, which satisfy $\langle \delta(\mathbf{r}, t) \rangle = \langle \tilde{\delta}(\mathbf{r}, t) \rangle = 0$, describe the atoms scattered into the collision halo. Note that here we have also implicitly used the property that $\Psi(\mathbf{r}, t)$ and $\tilde{\Psi}(\mathbf{r}, t)$ are related by conjugation only when taking the average of the stochastic fields.

Equations of motion for the fluctuating operators and mean-field components can be found by substitution of $\Psi(\mathbf{r}, t) \equiv \psi(\mathbf{r}, t) + \delta(\mathbf{r}, t)$ and $\Psi(\tilde{\mathbf{r}}, t) \equiv \psi(\mathbf{r}, t)^* + \tilde{\delta}(\mathbf{r}, t)$ back into Eqs. (1.108) and (1.109). The mean-field component $\psi(\mathbf{r}, t)$ evolves according to [31, 76]:

$$\frac{d\psi(\mathbf{r}, t)}{dt} = i\frac{\hbar}{2m}\nabla^2\psi(\mathbf{r}, t) - i\frac{U}{\hbar}|\psi(\mathbf{r}, t)|^2\psi(\mathbf{r}, t), \tag{1.112}$$

which corresponds to the standard time-dependent Gross–Pitaevskii equation. Similarly, we identify that, to lowest order, the fluctuating fields are described by:

$$\frac{d\delta(\mathbf{r}, t)}{dt} = i\frac{\hbar}{2m}\nabla^2\delta(\mathbf{r}, t) - 2i\frac{U}{\hbar}|\psi(\mathbf{r}, t)|^2[\tilde{\delta}(\mathbf{r}, t)]^* + \sqrt{-i\frac{U}{\hbar}}\psi(\mathbf{r}, t)\xi_1(\mathbf{r}, t), \tag{1.113}$$

$$\frac{d\tilde{\delta}(\mathbf{r}, t)}{dt} = -i\frac{\hbar}{2m}\nabla^2\tilde{\delta}(\mathbf{r}, t) + 2i\frac{U}{\hbar}|\psi(\mathbf{r}, t)|^2[\delta(\mathbf{r}, t)]^* + \sqrt{i\frac{U}{\hbar}}\psi(\mathbf{r}, t)^*\xi_2(\mathbf{r}, t), \tag{1.114}$$

The initial condition for the mean-field component is taken to be the same as for $\Psi(\mathbf{r}, 0)$ previously, $\psi(\mathbf{r}, t) = \sqrt{\rho(\mathbf{r}, 0)/2}e^{i\mathbf{k}_0 \cdot \mathbf{r}} + \sqrt{\rho(\mathbf{r}, 0)/2}e^{-i\mathbf{k}_0 \cdot \mathbf{r}}$, whereas the fluctuating component describes an initial vacuum state with $\delta(\mathbf{r}, t) = \tilde{\delta}(\mathbf{r}, t) = 0$ [64]. From inspection of these linearized equations for the stochastic fields we can see that rather than the highly unstable multiplicative noise in the full positive-P treatment [Eqs. (1.108) and (1.109)], the noise is now only additive and thus we may naively expect more stable stochastic trajectories.

These equations can be equivalently derived by considering a Bogoliubov approximation for the full quantum field operator, $\hat{\psi}(\mathbf{r}, t) = \psi(\mathbf{r}, t) + \hat{\delta}$, as outlined in Sect. 1.4.2, where again the mean-field term $\psi(\mathbf{r}, t) \equiv \langle \hat{\psi}(\mathbf{r}, t) \rangle$ describes the counter-propagating source condensates treated in the undepleted pump approximation, and $\hat{\delta}$ the atoms scattered into the collision halo. From inspection of Eqs. (1.113) and (1.114) an effective Hamiltonian describing the process is is given by:

$$\hat{H}_{\text{eff}} = \int d^3\mathbf{r} \left\{ \hat{\delta}^\dagger(\mathbf{r}, t) \left[-\frac{\hbar^2}{2m} \nabla^2 \right] \hat{\delta}(\mathbf{r}, t) + 2U \, |\psi(\mathbf{r}, t)|^2 \, \hat{\delta}^\dagger(\mathbf{r}, t) \hat{\delta}(\mathbf{r}, t) \right.$$

$$+ U \left[\psi_{+\mathbf{k}_0}(\mathbf{r}, t)\psi_{-\mathbf{k}_0}(\mathbf{r}, t)\hat{\delta}^\dagger(\mathbf{r}, t)\hat{\delta}^\dagger(\mathbf{r}, t) \right.$$

$$\left. \left. + \psi^*_{+\mathbf{k}_0}(\mathbf{r}, t)\psi^*_{-\mathbf{k}_0}(\mathbf{r}, t)\hat{\delta}(\mathbf{r}, t)\hat{\delta}(\mathbf{r}, t) \right] \right\}, \tag{1.115}$$

where the mean-field term $\psi(\mathbf{r}, t) = \psi_{+\mathbf{k}_0}(\mathbf{r}, t) + \psi_{-\mathbf{k}_0}(\mathbf{r}, t)$ is split further into the two counter-propagating condensates with momenta $\pm \mathbf{k}_0$ which are treated in the undepleted pump approximation. The quadratic nature of the effective Hamiltonian $\hat{H}_{\text{eff}}$ means that the boundary terms of the P distribution are guaranteed to vanish [61, 74] and thus the stochastic equations for the fields $\delta(\mathbf{r}, t)$ and $\tilde{\delta}(\mathbf{r}, t)$ [corresponding to $\hat{\delta}(\mathbf{r}, t)$ and $\hat{\delta}^\dagger(\mathbf{r}, t)$ respectively] will be stable in this sense. Hence, rather than a limited useful simulation duration, the stochastic Bogoliubov method is effectively limited by the validity of treating the source condensate in the undepleted pump approximation. This approximation is generally valid as long as the total number of scattered atoms is less than $\sim 10\%$ of the original source condensate population.

References

1. Einstein, A., Podolsky, B., Rosen, N.: Can quantum-mechanical description of physical reality be considered complete? Phys. Rev. **47**, 777–780 (1935)
2. Mermin, N.D.: What's wrong with this pillow? Phys. Today **42**, 9–11 (1989)
3. Bohm, D., Schiller, R., Tiomno, J.: A causal interpretation of the Pauli equation (a). Il Nuovo Cimento Ser. **10**(1), 48–66 (1955)
4. Ou, Z.Y., Pereira, S.F., Kimble, H.J., Peng, K.C.: Realization of the Einstein-Podolsky-Rosen paradox for continuous variables. Phys. Rev. Lett. **68**, 3663–3666 (1992)
5. Reid, M.D.: Demonstration of the Einstein-Podolsky-Rosen paradox using nondegenerate parametric amplification. Phys. Rev. A **40**, 913–923 (1989)
6. Braunstein, S.L., van Loock, P.: Quantum information with continuous variables. Rev. Mod. Phys. **77**, 513–577 (2005)
7. Gross, C., et al.: Atomic homodyne detection of continuous-variable entangled twin-atom states. Nature **480**, 219 (2011)
8. Raymer, M.G., Funk, A.C., Sanders, B.C., de Guise, H.: Separability criterion for separate quantum systems. Phys. Rev. A **67**, 052104 (2003)
9. Bell, J.S.: On the Einstein-Podolsky-Rosen paradox. Phys. (N.Y.) **1**, 195 (1964)
10. Aspect, A., Dalibard, J., Roger, G.: Experimental test of Bell's inequalities using time-varying analyzers. Phys. Rev. Lett. **49**, 1804–1807 (1982)
11. Weihs, G., Jennewein, T., Simon, C., Weinfurter, H., Zeilinger, A.: Violation of Bell's inequality under strict Einstein locality conditions. Phys. Rev. Lett. **81**, 5039–5043 (1998)
12. Rarity, J.G., Tapster, P.R.: Experimental violation of Bell's inequality based on phase and momentum. Phys. Rev. Lett. **64**, 2495–2498 (1990)
13. Rowe, M.A., et al.: Experimental violation of a Bell's inequality with efficient detection. Nature **409**, 791–794 (2001)
14. Sakai, H., et al.: Spin correlations of strongly interacting massive fermion pairs as a test of Bell's inequality. Phys. Rev. Lett. **97**, 150405 (2006)
15. Garg, A., Mermin, N.D.: Detector inefficiencies in the Einstein-Podolsky-Rosen experiment. Phys. Rev. D **35**, 3831–3835 (1987)

16. Zurek, W.H.: Decoherence and the transition from quantum to classical–revisited. arXiv preprint (2003). arXiv:quant-ph/0306072
17. Penrose, R.: On gravity's role in quantum state reduction. Gen. Relativ. Gravit. **28**, 581–600 (1996)
18. Yurke, B., McCall, S.L., Klauder, J.R.: SU(2) and SU(1,1) interferometers. Phys. Rev. A **33**, 4033–4054 (1986)
19. Walls, D.F., Milburn, G.J.: Quantum Optics, 2nd edn. Springer, Berlin (2008)
20. Glauber, R.J.: The quantum theory of optical coherence. Phys. Rev. Lett. **130**, 2529–2539 (1963)
21. Hanbury Brown, R., Twiss, H.Q.: Interferometry of the intensity fluctuations in light, i. basic theory: The correlation between photons in coherent beams of radiation. Proc. R. Soc. Lond. **242**, 300–324 (1957)
22. Su, C., Wódkiewicz, K.: Quantum versus stochastic or hidden-variable fluctuations in two-photon interference effects. Phys. Rev. A **44**, 6097–6108 (1991)
23. Walls, D.F., Milburn, G.: Quantum Optics. Springer Study Edition. Springer, Berlin (1995)
24. Reid, M.D., Walls, D.F.: Violations of classical inequalities in quantum optics. Phys. Rev. A **34**, 1260–1276 (1986)
25. Perrin, A., et al.: Observation of atom pairs in spontaneous four-wave mixing of two colliding Bose-Einstein condensates. Phys. Rev. Lett. **99**, 150405 (2007)
26. Perrin, A., et al.: Atomic four-wave mixing via condensate collisions. New J. Phys. **10**, 045021 (2008)
27. Mølmer, K., et al.: Hanbury Brown and Twiss correlations in atoms scattered from colliding condensates. Phys. Rev. A **77**, 033601 (2008)
28. Kheruntsyan, K.V., et al.: Violation of the Cauchy-Schwarz inequality with matter waves. Phys. Rev. Lett. **108**, 260401 (2012)
29. Jaskula, J.-C., et al.: Sub-Poissonian number differences in four-wave mixing of matter waves. Phys. Rev. Lett. **105**, 190402 (2010)
30. Jeltes, T., et al.: Comparison of the Hanbury Brown-Twiss effect for bosons and fermions. Nature **445**, 402–405 (2007)
31. Krachmalnicoff, V., et al.: Spontaneous four-wave mixing of de Broglie waves: Beyond optics. Phys. Rev. Lett. **104**, 150402 (2010)
32. Deuar, P., et al.: Anisotropy in s-wave Bose-Einstein condensate collisions and its relationship to superradiance. Phys. Rev. A **90**, 033613 (2014)
33. Chwedeńczuk, J., et al.: Pair correlations of scattered atoms from two colliding Bose-Einstein condensates: Perturbative approach. Phys. Rev. A **78**, 053605 (2008)
34. Ogren, M., Kheruntsyan, K.V.: Atom-atom correlations in colliding Bose-Einstein condensates. Phys. Rev. A. **79**, 021606 (2009)
35. Ziń, P., Chwedeńczuk, J., Trippenbach, M.: Elastic scattering losses from colliding Bose-Einstein condensates. Phys. Rev. A **73**, 033602 (2006)
36. Ohmi, T., Machida, K.: Bose-Einstein condensation with internal degrees of freedom in alkali atom gases. J. Phys. Soc. Jpn. **67**, 1822–1825 (1998)
37. Saito, H., Kawaguchi, Y., Ueda, M.: Breaking of chiral symmetry and spontaneous rotation in a spinor Bose-Einstein condensate. Phys. Rev. Lett. **96**, 065302 (2006)
38. Gross, C., et al.: Squeezing and entanglement in a Bose-Einstein condensate. Nature **480**, 219–223 (2011)
39. Law, C.K., Pu, H., Bigelow, N.P.: Quantum spins mixing in spinor Bose-Einstein condensates. Phys. Rev. Lett. **81**, 5257–5261 (1998)
40. Zhang, W., Zhou, D.L., Chang, M.-S., Chapman, M.S., You, L.: Coherent spin mixing dynamics in a spin-1 atomic condensate. Phys. Rev. A **72**, 013602 (2005)
41. Zhang, W., You, L.: An effective quasi-one-dimensional description of a spin-1 atomic condensate. Phys. Rev. A **71**, 025603 (2005)
42. Pu, H., Law, C.K., Raghavan, S., Eberly, J.H., Bigelow, N.P.: Spin-mixing dynamics of a spinor Bose-Einstein condensate. Phys. Rev. A **60**, 1463–1470 (1999)

43. Chang, M.-S., et al.: Observation of spinor dynamics in optically trapped ^{87}Rb Bose-Einstein condensates. Phys. Rev. Lett. **92**, 140403 (2004)
44. Chang, M.-S., Qin, Q., Zhang, W., You, L., Chapman, M.S.: Coherent spinor dynamics in a spin-1 Bose condensate. Nat. Phys. **1**, 111–116 (2005)
45. Kronjäger, J., et al.: Evolution of a spinor condensate: Coherent dynamics, dephasing, and revivals. Phys. Rev. A **72**, 063619 (2005)
46. Klempt, C., et al.: Multiresonant spinor dynamics in a Bose-Einstein condensate. Phys. Rev. Lett. **103**, 195302 (2009)
47. Klempt, C., et al.: Parametric amplification of vacuum fluctuations in a spinor condensate. Phys. Rev. Lett. **104**, 195303 (2010)
48. Hoang, T.M., et al.: Dynamic stabilization of a quantum many-body spin system. Phys. Rev. Lett. **111**, 090403 (2013)
49. Oberthaler, M.K., Linnemann, D.: Private Communication (2015)
50. Linnemann, D., Lewis-Swan, R.J., Strobel, H., Mussel, W., Kheruntstyan, K.V., Oberthaler, M.K.: Quantum-enhanced sensing based on time reversal of non-linear dynamics (2016). arXiv:1602.07505
51. Kawaguchi, Y., Ueda, M.: Spinor Bose-Einstein condensates. Phys. Rep. **520**, 253–381 (2012)
52. Stamper-Kurn, D.M., Ueda, M.: Spinor Bose gases: Symmetries, magnetism, and quantum dynamics. Rev. Mod. Phys. **85**, 1191–1244 (2013)
53. Widera, A., et al.: Precision measurement of spin-dependent interaction strengths for spin-1 and spin-2 87 rb atoms. New J. Phys. **8**, 152 (2006)
54. Schmaljohann, H., et al.: Dynamics of $F = 2$ spinor Bose-Einstein condensates. Phys. Rev. Lett. **92**, 040402 (2004)
55. Lewis-Swan, R.J., Kheruntsyan, K.V.: Sensitivity to thermal noise of atomic Einstein-Podolsky-Rosen entanglement. Phys. Rev. A **87**, 063635 (2013)
56. Wigner, E.: On the quantum correction for thermodynamic equilibrium. Phys. Rev. **40**, 749–759 (1932)
57. Polkovnikov, A.: Phase space representation of quantum dynamics. Ann. Phys. **325**, 1790–1852 (2010)
58. Schleich, W.P.: Quantum Optics in Phase Space. Wiley, Berlin (2011)
59. Leonhardt, U.: Essential Quantum Optics. Cambridge University Press, Cambridge (2010)
60. Opanchuk, B.: Quasiprobability methods in quantum interferometry of ultracold matter. Ph.D. thesis, Swinburne University of Technology (2014)
61. Gardiner, C.W.: Handbook of Stochastic Methods, vol. 4. Springer, Berlin (1985)
62. Gardiner, C. & Zoller, P. *Quantum noise*, vol. 56 (Springer Science & Business Media, 2004)
63. Sinatra, A., Lobo, C., Castin, Y.: The truncated Wigner method for Bose-condensed gases: limits of validity and applications. J. Phys. B **35**, 3599 (2002)
64. Olsen, M.K., Bradley, A.S.: Numerical representation of quantum states in the positive-P and Wigner representations. Opt. Comm. **282**, 3924–3929 (2009)
65. Ruostekoski, J. & Martin, A. Truncated wigner method for bose gases. In Gardiner, S.A., Proukakis, N., Davis, M.J. & Szymanska, M. (eds.) *Quantum gases: Finite temperature and non-equilibrium dynamics* (World Scientific, 2013)
66. Isella, L., Ruostekoski, J.: Nonadiabatic dynamics of a bose-einstein condensate in an optical lattice. Phys. Rev. A **72**, 011601 (2005)
67. Barnett, R., Polkovnikov, A., Vengalattore, M.: Prethermalization in quenched spinor condensates. Phys. Rev. A **84**, 023606 (2011)
68. Drummond, P., Gardiner, C.: Generalised P-representations in quantum optics. J. Phys. A **13**, 2353 (1980)
69. Vaughan, T.G.: The quantum dynamics of dilute gas BEC formation. Master's thesis, University of Queensland (2001)
70. Gardiner, C., Drummond, P.: Ten years of the positive P-representation. In: Inguva, R. (ed.) Recent Developments in Quantum Optics, pp. 77–86. Springer, Berlin (1993)
71. Dennis, G.R., Hope, J.J., Johnsson, M.T.: Xmds2: Fast, scalable simulation of coupled stochastic partial differential equations. Comput. Phys. Commun. **184**, 201–208 (2013)

72. Abrikosov, A., Gorkov, L., Dzyaloshinski, I., Dzialoshinskiui, I.: Methods of Quantum Field Theory in Statistical Physics. Dover Books on Physics Series. Dover Publications, New York (1975)
73. Deuar, P., Drummond, P.D.: First-principles quantum dynamics in interacting Bose gases: I. the positive P representation. J. Phys. A: Math. Gen. **39**, 1163 (2006)
74. Gilchrist, A., Gardiner, C.W., Drummond, P.D.: Positive P representation: Application and validity. Phys. Rev. A **55**, 3014–3032 (1997)
75. Ögren, M., Kheruntsyan, K.V., Corney, J.F.: First-principles quantum dynamics for fermions: Application to molecular dissociation. EPL **92**, 36003 (2010)
76. Deuar, P., Chwedeńczuk, J., Trippenbach, M., Ziń, P.: Bogoliubov dynamics of condensate collisions using the positive-P representation. Phys. Rev. A **83**, 063625 (2011)

Chapter 2
Proposal for Demonstrating the Hong–Ou–Mandel Effect with Matter Waves

Two-particle interference is a quintessential effect of quantum mechanics which is perhaps most beautifully demonstrated by the Hong–Ou–Mandel effect. In this phenomenon, the probability amplitudes of two indistinguishable photons entering opposing inputs of a beam-splitter interfere destructively, in a manner which is not describable by any classical theory. When realized with photons prepared in the two-mode squeezed vacuum state [1], this two-particle interference also serves as a demonstration of the strong non-classical correlations between the modes, in particular a violation of the Cauchy–Schwarz inequality. This elegant effect is thus intrinsically related to a violation of a Bell inequality, as both phenomena rely on underlying non-classical features of the quantum state.

In this chapter we outline a proposal to demonstrate the effect with massive particles, utilizing pairs of atoms produced by spontaneous four-wave mixing via colliding condensates, which, as demonstrated in Sect. 1.4.2, reduces in the simplest model to the same two-mode squeezed vacuum state. However, unlike the two-mode quantum optics scheme, the multimode nature of the collision halo motivates us to formulate a new measurement protocol to quantify the effect in the atomic case. An experimental demonstration of the effect has a two-fold impact for future tests of a Bell inequality in this system. Firstly, the interferometric scheme required for the Hong–Ou–Mandel effect, comprising of a series of laser-induced Bragg pulses (the atom-optics analogs of mirrors and beam-splitters), is strongly related to the Rarity–Tapster setup employed in Chap. 3 and thus acts as a stepping-stone for any experimental proposal involving atom-optics mirrors and beam-splitters. Secondly, as discussed above, a true demonstration of the effect requires an interference visibility of more than 50 % (relative to the background level of distinguishable paths through the beam-splitter) which is equivalent to a violation of the classical Cauchy–Schwarz inequality. Such non-classical correlations are a pre-requisite for a violation of a Bell inequality.

© Springer International Publishing Switzerland 2016
R.J. Lewis-Swan, *Ultracold Atoms for Foundational Tests of Quantum Mechanics*, Springer Theses, DOI 10.1007/978-3-319-41048-7_2

The remainder of this chapter is adapted from the published article: *'Proposal for demonstrating the Hong–Ou–Mandel effect with matter waves'* [R.J. Lewis-Swan and K.V. Kheruntsyan, Nature Comm. **5**, 3752 (2013)]. The supplementary information of this article can be found in Appendix C.

2.1 Introduction

Since its first demonstration, the Hong–Ou–Mandel (HOM) effect [1] has become a textbook example of quantum mechanical two-particle interference using pairs of indistinguishable photons. When two such photons enter a 50:50 beam splitter, with one photon in each input port, they both preferentially exit from the same output port, even though each photon individually had a 50:50 chance of exiting through either output port. The HOM effect was first demonstrated using optical parametric down-conversion [1]; the same setup, but with an addition of linear polarisers, was subsequently used to demonstrate a violation of a Bell inequality [2] which is of fundamental importance to validating some of the foundational principles of quantum mechanics such as quantum nonlocality and long-distance entanglement.

The HOM effect is a result of destructive quantum interference in a (bosonic) twin-photon state, which leads to a characteristic dip in the photon coincidence counts at two photodetectors placed at the output ports of a beam splitter. The destructive interference occurs between two *indistinguishable* paths corresponding to the photons being both reflected from, or both transmitted through, the beam splitter. Apart from being of fundamental importance to quantum physics, the HOM effect underlies the basic entangling mechanism in linear optical quantum computing [3], in which a twin-photon state $|1, 1\rangle$ is converted into a quantum superposition $\frac{1}{\sqrt{2}}(|2, 0\rangle - |0, 2\rangle)$—the simplest example of the elusive 'NOON' state [4]. Whereas the HOM effect with (massless) photons has been extensively studied in quantum optics (see [5, 6] and references therein), two-particle quantum interference with massive particles remains largely unexplored. A matter wave demonstration of the HOM effect would be a major advance in experimental quantum physics, enabling an expansion of foundational tests of quantum mechanics into previously unexplored regimes.

Here we propose an experiment which can realise the HOM effect with matter waves using a collision of two atomic Bose–Einstein condensates (BECs) (as in Refs. [7–11]) and a sequence of laser-induced Bragg pulses. The HOM interferometer uses pair-correlated atoms from the scattering halo that is generated during the collision through the process of spontaneous four-wave mixing. The pair-correlated atoms are being mixed with a sequence of two Bragg pulses [12, 13] in analogy with the use of twin-photons from parametric down conversion in the optical HOM scheme. The HOM effect is quantified via the measurement of a set of atom-atom pair correlation functions between the output ports of the interferometer. Using stochastic quantum simulations of the collisional dynamics and the sequence of Bragg pulses, we predict a HOM-dip visibility of ∼69% for realistic experimental parameters. A visibility larger than ∼50% is indicative of stronger than classical correlations between the atoms in the scattering halo [10, 11, 14–16], which in turn renders our system as a suitable platform for demonstrating a Bell's inequality violation with matter waves using a closely related Rarity–Tapster scheme [17].

2.2 Setup

The schematic diagram of the proposed experiment is shown in Fig. 2.1. A highly elongated (along the x-axis) BEC is initially split into two equal and counterpropagating halves traveling with momenta $\pm\mathbf{k}_0$ along z in the centre-of-mass frame. Constituent atoms undergo binary elastic collisions which produce a nearly spherical s-wave scattering halo of radius $k_r \simeq 0.95|\mathbf{k}_0|$ [9] in momentum space due to energy and momentum conservation. The elongated condensates have a disk shaped density distribution in momentum space, shown in Fig. 2.1b on the north and south poles of the halo. After the end of the collision (which in this geometry corresponds to complete spatial separation of the condensates in position space) we apply two counterpropagating lasers along the x-axis whose intensity and frequency are tuned to act as a resonant Bragg π-pulse with respect to two diametrically opposing momentum modes, $\mathbf{k}_1$ and $\mathbf{k}_2 = -\mathbf{k}_1$, situated on the equatorial plane of the halo and satisfying $|\mathbf{k}_{1,2}| = k_r$.

Previous experiments and theoretical work [7, 8, 10, 11, 18–23] have shown the existence of strong atom-atom correlation between such diametrically opposite modes, similar to the correlation between twin-photons in parametric down conver-

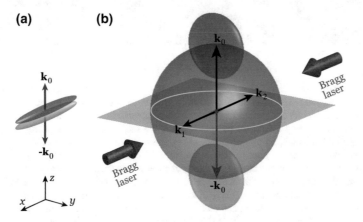

Fig. 2.1 **a** Schematic diagram of the geometry of collision of two elongated Bose–Einstein condensates in position space. **b** Momentum space distribution of the atomic cloud showing the (*disk shaped*) colliding condensates on the north and south poles of the spherical halo of scattered atoms (see text for further details)

sion. Applying the Bragg π-pulse to the collisional halo replicates an optical mirror and reverses the trajectories of the scattered atoms with momenta $\mathbf{k}_1$ and $\mathbf{k}_2$, and a finite region around them. We assume that the pulse is tuned to operate in the so-called Bragg regime of the Kapitza–Dirac effect [13, 24] (diffraction of a matter-wave from a standing light field), corresponding to conditions in which second- and higher-order diffractions are suppressed. The system is then allowed to propagate freely for a duration so that the targeted atomic wave-packets regain spatial overlap in position space. We then apply a second Bragg pulse—a $\pi/2$-pulse—to replicate an optical 50:50 beam-splitter, which is again targeted to couple $\mathbf{k}_1$ and $\mathbf{k}_2$, thus realising the HOM interferometer.

The timeline of the proposed experiment is illustrated in Fig. 2.2a, whereas the results of numerical simulations (see Methods) of the collision dynamics and the application of Bragg pulses are shown in Fig. 2.2b–d: (b) shows the equatorial slice of the momentum-space density distribution $n(\mathbf{k}, t)$ of the scattering halo at the end of collision; (c) and (d) show the halo density after the application of the π and $\pi/2$ pulses, respectively. The 'banana' shaped regions in (c) correspond to 'kicked' populations between the targeted momenta around $\mathbf{k}_1$ and $\mathbf{k}_2$ in the original scattering

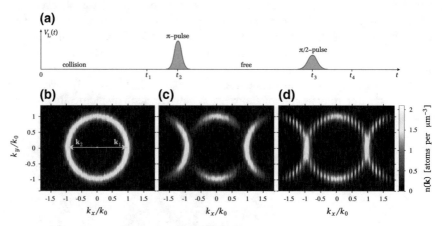

Fig. 2.2 **a**—Timeline of the proposed experiment; **b–d**—the results of numerical simulations showing the momentum-space density distribution $n(\mathbf{k})$ of scattered atoms on the equatorial plane of the halo. In panel (**a**), $V_L(t)$ denotes the depth of the lattice potential formed by the Bragg lasers, with the first hump indicating the mirror (π) pulse, while the second hump—the beam-splitter ($\pi/2$) pulse (the initial source-splitting pulse that sets up the collision of condensates is not shown for clarity). Panel (**b**) shows the density distribution after the collision, at $t_1 = 65\,\mu s$; **c**—after the π-pulse, centred at $t_2 = 75\,\mu s$ and having a duration of $\tau_\pi = 2.5\,\mu s$ (rms width of Gaussian envelope); and **d**—after the final $\pi/2$ pulse, with $\Delta t_{\text{free}} = t_3 - t_2 = 85\,\mu s$ and $\tau_{\pi/2} = 2.5\,\mu s$ (see Methods for further details; the durations shown on the time axis are not to scale). The momentum axes $k_{x,y}$ in panels (**b**)–(**d**) are normalised to the collision momentum $k_0 \equiv |\mathbf{k}_0|$ (in wave-number units), which in our simulations was $k_0 = 4.7 \times 10^6\,\text{m}^{-1}$. The simulations were carried out for an initial BEC containing a total of $N = 4.7 \times 10^4$ atoms of metastable helium ($^4\text{He}^*$), prepared in a harmonic trap of frequencies $(\omega_x, \omega_y, \omega_z)/2\pi = (64, 1150, 1150)\,\text{Hz}$, and colliding with the scattering length of $a = 5.3\,\text{nm}$; all these parameters are very close to those realised in recent experiments [9, 10]

halo, while (d) shows the density distribution after mixing. The density modulation in (c) is simply the result of interference between the residual and transferred atomic populations after the π-pulse upon their recombination on the beamsplitter. The residual population is due to the fact that the pairs of off-resonance modes in these parts of the halo (which are coupled by the same Bragg pulses as they share the same momentum difference $2k_r$ as the resonant modes $\mathbf{k}_1$ and $\mathbf{k}_2$) no longer satisfy the perfect Bragg resonance condition and therefore the population transfer during the π-pulse is not 100 % efficient (see Supplementary Information). As these components have unequal absolute momenta, their amplitudes accumulate a nonzero relative phase due to phase dispersion during the free propagation. The accrued relative phase results in interference fringes upon the recombination on the beamsplitter, with an approximate period of $\Delta k \simeq \pi m/(\hbar k_r \Delta t_{\text{free}}) \simeq 0.1 |\mathbf{k}_0|$.

Due to the indistinguishability of the paths of the Bragg-resonant modes $\mathbf{k}_1$ and $\mathbf{k}_2$ through the beam-splitter and the resulting destructive quantum interference, a measurement of coincidence counts between the atomic populations in these modes will reveal a suppression compared to the background level. To reveal the full structure of the HOM dip, including the background level where no quantum interference occurs, we must introduce path distinguishability between the $\mathbf{k}_1$ and $\mathbf{k}_2$ modes. One way to achieve this, which would be in a direct analogy with shifting the beam splitter in the optical HOM scheme, is to change the Bragg-pulse resonance condition from the $(\mathbf{k}_1, \mathbf{k}_2)$ pair to $(\mathbf{k}_1, \mathbf{k}_2 + \hat{\mathbf{e}}_x \delta k)$, where $\hat{\mathbf{e}}_x$ is the unit vector in the x-direction. The approach to the background coincidence rate between the populations in the $\mathbf{k}_1$ and $\mathbf{k}_2$ modes would then correspond to performing the same experiment for increasingly large displacements δk. Taking into account that acquiring statistically significant results for each δk requires repeated runs of the experiment (typically thousands), this measurement protocol could potentially pose a significant practical challenge due to the very large total number of experimental runs required.

2.3 Results and Discussion

To overcome this challenge, we propose an alternative measurement protocol which can reveal the full structure of the HOM dip from just one Bragg-resonance condition, requiring only one set of experimental runs. The protocol takes advantage of the broadband, multimode nature of the scattering halo and the fact that the original Bragg pulse couples not only the targeted momentum modes $\mathbf{k}_1$ and $\mathbf{k}_2$, but also many other pairs of modes which follow distinguishable paths through the beam-splitter. One such pair, $\mathbf{k}_3 = (k_x, k_y, k_z) = k_r(\cos(\theta), \sin(\theta), 0)$ and $\mathbf{k}_4 = -\mathbf{k}_3$, located on the halo peak, is shown in Fig. 2.3a and corresponds to a rotation by angle θ away from $\mathbf{k}_1$ and $\mathbf{k}_2$. The modes $\mathbf{k}_3$ and $\mathbf{k}_4$ are equivalent to the original pair in the sense of their quantum statistical properties and therefore, these modes can be used for the measurement of the background level of coincidence counts, instead of physically altering the paths of the $\mathbf{k}_1$ and $\mathbf{k}_2$ modes. The angle θ now serves the role of the 'displacement' parameter that scans through the shape of the HOM dip.

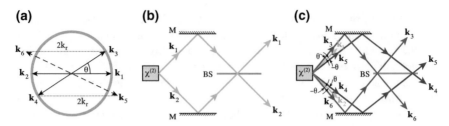

Fig. 2.3 Panel (**a**) shows the schematic of a set of momentum modes affected by the Bragg pulses. The diametrically opposite vectors $\mathbf{k}_1$ and $\mathbf{k}_2 = -\mathbf{k}_1$ show the targeted modes; their amplitude is given by the halo peak radius, $k_r = |\mathbf{k}_1| = |\mathbf{k}_2|$, which is equal to $k_r = 0.95|\mathbf{k}_0|$ in this part of the halo [9]. Also shown are the to-be-measured momentum components $\mathbf{k}_3$ and $\mathbf{k}_4$ corresponding to a rotation by θ away from the targeted modes, which couple, respectively, to $\mathbf{k}_6 = \mathbf{k}_3 - 2\mathbf{k}_1$ and $\mathbf{k}_5 = \mathbf{k}_4 + 2\mathbf{k}_1$ by the same Bragg pulses. Panels (**b**) and (**c**) show a topologically equivalent optical scheme. A $\chi^{(2)}$ nonlinear crystal is optically pumped to produce twin-photons via parametric down-conversion. In (**b**) we depict the archetypal optical HOM setup which corresponds to the case of $\theta = 0$ in the atom-optics scheme. A twin-photon state in modes $\mathbf{k}_1$ and $\mathbf{k}_2$ is first selected from a broadband source, then mixed at the beam-splitter (*BS*) after reflection from the mirror (*M*), and photon coincidence counts are measured between the two symmetric output ports of the interferometer. In (**c**) we depict the optical setup which is equivalent to $\theta > 0$ in the atom-optics proposal. Two twin-photon states in ($\mathbf{k}_3$, $\mathbf{k}_4$) and in ($\mathbf{k}_5$, $\mathbf{k}_6$) are selected from the broadband source; the asymmetry of the pairs about the optical axis of the interferometer means that the correlated photons from the respective pairs will arrive at the beam-splitter at spatially separate locations and will mix with photons from the other pair, which introduces distinguishability between the paths through the interferometer

A topologically equivalent optical scheme is shown in Fig. 2.3b, c, which is in turn similar to the one analysed in Ref. [25] using a broadband source of angle-separated pair-photons and directionally asymmetric apertures.

In the proposed protocol, detection (after the final Bragg pulse) of atom coincidences at the pair of originally correlated momenta $\mathbf{k}_3$ and $\mathbf{k}_4$ corresponds to both paths being separately *reflected* on the beamsplitter (see Fig. 2.3c). Apart from this outcome, we need to take into account the coincidences between the respective Bragg-partner momenta, $\mathbf{k}_6$ and $\mathbf{k}_5$ (separated, respectively, from $\mathbf{k}_3$ and $\mathbf{k}_4$ by the same difference $2k_r$ as $\mathbf{k}_1$ from $\mathbf{k}_2$). Coincidences at $\mathbf{k}_6$ and $\mathbf{k}_5$ correspond to atoms of the originally correlated momenta $\mathbf{k}_3$ and $\mathbf{k}_4$ being both *transmitted* through the beam splitter (see Fig. 2.3c). Finally, in order to take into account all possible channels contributing to coincidence counts between the two arms of the interferometer, we need to measure coincidences between $\mathbf{k}_3$ and $\mathbf{k}_6$, as well as between $\mathbf{k}_4$ and $\mathbf{k}_5$. This ensures that the total detected flux at the output ports of the beam splitter matches the total input flux. In addition to this, we normalise the bare coincidence counts to the product of single-detector count rates, i.e., the product of the average number of atoms in the two output arms of the interferometer. We use the normalised correlation function as the total population in the four relevant modes varies as the angle θ is increased, implying that the raw coincidence rates are not a suitable quantity to compare at different angles.

With this measurement protocol in mind, we quantify the HOM effect using the normalised second-order correlation function $\overline{g}^{(2)}_{RL}(t) = \langle:\hat{N}_R(t)\hat{N}_L(t):\rangle/\langle\hat{N}_R(t)\rangle$ $\langle\hat{N}_L(t)\rangle$ after the $\pi/2$-pulse concludes at $t = t_4$. Here, $\langle\hat{N}_R\rangle \equiv \langle\hat{N}_3\rangle + \langle\hat{N}_5\rangle$ and $\langle\hat{N}_L\rangle \equiv \langle\hat{N}_4\rangle + \langle\hat{N}_6\rangle$ correspond to the number of atoms detected, respectively, on the two (right and left) output ports of the beam splitter, with the detection bins centred around the four momenta of interest $\mathbf{k}_i$ ($i = 3, 4, 5$, and 6), for any given angle θ [see Fig. 2.2e]. More specifically, $\hat{N}_i(t) = \int_{\mathcal{V}(\mathbf{k}_i)} d^3\mathbf{k}\,\hat{n}(\mathbf{k}, t)$ is the atom number operator in the integration volume $\mathcal{V}(\mathbf{k}_i)$ centred around $\mathbf{k}_i$, where $\hat{n}(\mathbf{k}, t) = \hat{a}^\dagger(\mathbf{k}, t)\hat{a}(\mathbf{k}, t)$ is the momentum-space density operator, with $\hat{a}^\dagger(\mathbf{k}, t)$ and $\hat{a}(\mathbf{k}, t)$ the corresponding creation and annihilation operators (the Fourier components of the field operators $\hat{\delta}^\dagger(\mathbf{r}, t)$ and $\hat{\delta}(\mathbf{r}, t)$, see Methods). The double-colon notation in $\langle:\hat{N}_R(t)\hat{N}_L(t):\rangle$ indicates normal ordering of the creation and annihilation operators.

The integrated form of the second-order correlation function, which quantifies the correlations in terms of atom number coincidences in detection bins of certain size rather than in terms of local density-density correlations, accounts for limitations in the experimental detector resolution, in addition to improving the signal-to-noise ratio which is typically low due to the relatively low density of the scattering halo; in typical condensate collision experiments and in our simulations, the low density translates to a typical halo-mode occupation of $\sim$0.1. We choose $\mathcal{V}(\mathbf{k}_i)$ to be a rectangular box with dimensions corresponding to the rms width of the initial momentum distribution of the trapped condensate, which is a reasonable approximation to the mode (or coherence) volume in the scattering halo [8, 22].

The second-order correlation function $\overline{g}^{(2)}_{RL}(t_4)$, quantifying the HOM effect as a function of the path-distinquishability angle θ, is shown in Fig. 2.4. For $\theta = 0$, where $\mathbf{k}_{3(4)} = \mathbf{k}_{1(2)}$, we observe maximum suppression of coincidence counts relative to the background level due to the indistinguishability of the paths. As we increase $|\theta| > 0$,

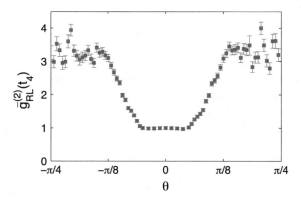

Fig. 2.4 Normalised atom-atom correlation function $\overline{g}^{(2)}_{RL}(t_4)$ between the two arms of the interferometer, characterising the HOM effect as a function of the path-distinguishability angle θ. Error bars denote sampling error from $\sim$30,000 stochastic simulations (see Methods). The atom counting bins are *rectangular boxes* with sides $\delta k_x = 0.01k_0$ and $\delta k_{y,z} = 0.19k_0$ which approximate the widths of the momentum distribution of the initial trapped BEC

we no longer mix $\mathbf{k}_3$ and $\mathbf{k}_4$ as a pair and their paths through the beam-splitter become distinguishable; the path interference is lost, and we observe an increase in the magnitude of the correlation function to the background level. We quantify the visibility of the HOM dip via $V = 1 - \min\{\overline{g}_{RL}^{(2)}(t_4)\} / \max\{\overline{g}_{RL}^{(2)}(t_4)\}$, where $\min\{\overline{g}_{RL}^{(2)}(t_4)\}$ occurs for $\theta = 0$ and $\max\{\overline{g}_{RL}^{(2)}(t_4)\}$ for sufficiently large θ such that momenta $\mathbf{k}_{5,6}$ lie outside the scattering halo. Due to the oscillatory nature of the wings (see below) we take $\max\{\overline{g}_{RL}^{(2)}(t_4)\}$ to correspond to the mean of $\overline{g}_{RL}^{(2)}(t_4)$ for $\theta \gtrsim \pi/8$. Using this definition we measure a visibility of $V \simeq 0.69 \pm 0.09$, where the uncertainty of ± 0.09 corresponds to taking into account the full fluctuations of $g_{RL}^{(2)}(t_4)$ about the mean in the wings rather than fitting the oscillations (see Supplementary Information). The visibility larger than 0.5 is consistent with the nonclassical effect of violation of Cauchy–Schwarz inequality with matter waves, observed recently in condensate collision experiments [11]. The exact relationship between the visibility and the Cauchy–Schwarz inequality is discussed further in the Supplementary Information, as are simple (approximate) analytic estimates of the magnitude of the HOM dip visibility.

The broadband, multimode nature of the scattering halo implies that the range of the path-length difference over which the HOM effect can be observed is determined by the spectral width of the density profile of the scattering halo. Therefore the width of the HOM dip is related to the width of the halo density. This is similar to the situation analysed in Ref. [25] using pair-photons from a broadband parametric down-converter. The angular width of the HOM dip extracted from Fig. 2.4 is approximately $w_{HOM} \simeq 0.61$ radians, which is indeed close to the width (full width at half maximum) of the scattering halo in the relevant direction, $w_{halo} \simeq 0.69$ radians (see also Supplementary Information for simple analytic estimates). The same multimode nature of the scattering halo contributes to the oscillatory behaviour in the wings of the HOM dip profile: here we mix halo modes with unequal absolute momenta and the resulting phase dispersion from free-propagation leads to oscillations similar to those observed with two-color photons [25].

We emphasise that the input state in our matter-wave HOM interferometer is subtly different from the idealised twin-Fock state $|1, 1\rangle$ used in the simplest analytic descriptions of the optical HOM effect. This idealised state stems from treating the process of spontaneous optical parametric down-conversion (SPDC) in the weak-gain regime. We illustrate this approximation by considering a two-mode toy model of the process, which in the undepleted pump approximation is described by the Hamiltonian $\hat{H} = \hbar g(\hat{a}_1^\dagger \hat{a}_2^\dagger + h.c.)$ that produces perfectly correlated photons in the $\hat{a}_1$ and $\hat{a}_2$ modes, where $g > 0$ is a gain coefficient related to the quadratic nonlinearity of the medium and the amplitude of the coherent pump beam. (In the context of condensate collisions, the coupling g corresponds to $g = U\rho_0(0)/\hbar$ at the same level of 'undepleted pump' approximation [8, 22]; see Methods for the definitions of U and ρ_0.) The full output state of the SPDC process in the Schrödinger picture is given by $|\psi\rangle = \sqrt{1 - \alpha^2} \sum_{n=0}^{\infty} \alpha^n |n, n\rangle$, where $\alpha = \tanh(gt)$ and t is the interaction time [26]. In the weak-gain regime, corresponding to $\alpha \simeq gt \ll 1$, this state is well approximated by $|\psi\rangle \propto |0, 0\rangle + \alpha|1, 1\rangle$, i.e., by truncating the expansion of $|\psi\rangle$

and neglecting the contribution of the $|2, 2\rangle$ and higher-n components. This regime corresponds to mode populations being much smaller than one, $\langle \hat{n} \rangle = \langle \hat{a}^{\dagger}_{1(2)} \hat{a}_{1(2)} \rangle = \sinh^2(gt) \simeq (gt)^2 \simeq \alpha^2 \ll 1$. The truncated state itself is qualitatively identical to the idealised state $|1, 1\rangle$ as an input to the HOM interferometer: both result in a HOM dip minimum of $\bar{g}^{(2)}_{RL} = 0$ and $\bar{g}^{(2)}_{RL} \simeq 1/2\langle \hat{n} \rangle$ in the wings, with the resulting maximum visibility of $V = 1$. If, on the other hand, the contribution of the $|2, 2\rangle$ and higher-n components is not negligible (which is the case, for example, of $\langle \hat{n} \rangle \simeq 0.1$) then the raw coincidence counts at the HOM dip and the respective normalised correlation function no longer equal to zero; in fact, the full SPDC state for arbitrary $\alpha < 1$ leads to a HOM dip minimum of $\bar{g}^{(2)}_{RL} = 1$ and $\bar{g}^{(2)}_{RL} = 2 + 1/2\langle \hat{n} \rangle$ in the wings, which in turn results in a reduced visibility of $V = 1 - 1/(2 + 1/2\langle \hat{n} \rangle)$.

The process of four-wave mixing of matter-waves gives rise to an output state analogous to the above SPDC state for each pair of correlated modes (see, e.g., [8, 22] and Supplementary Information). Indeed, the fraction of atoms converted from the source BEC to all scattering modes is typically less than 5 %, which justifies the use of the undepleted pump approximation. The typical occupation numbers of the scattered modes are, however, beyond the extreme of a very weak gain. In our simulations, the mode occupation on the scattering halo is on the order of 0.1 and therefore, even in the simplified analytical treatment of the process, the output state of any given pair of correlated modes cannot be approximated by the truncated state $|0, 0\rangle + \alpha|1, 1\rangle$ or indeed the idealised twin-Fock state $|1, 1\rangle$.

At the basic level, our proposal only relies on the existence of the aforementioned pair-correlations between scattered atoms, with the strength of the correlations affecting the visibility of the HOM dip. For a sufficiently homogeneous source BEC [22, 27], the correlations and thus the visibility V effectively depend only on the average mode population $\langle \hat{n} \rangle$ in the scattering halo, with a scaling of V on $\langle \hat{n} \rangle$ given by $V = 1 - 1/(2 + 1/2\langle \hat{n} \rangle)$ by our analytic model. Dependence of $\langle \hat{n} \rangle$ on system parameters such as the total number of atoms in the initial BEC, trap frequencies, and collision duration is well understood both theoretically and experimentally [8–11], and each can be sufficiently controlled such that a suitable mode population of $\langle \hat{n} \rangle \lesssim 1$ can, in principle, be targeted. There lies, however, a need for optimisation: very small populations are preferred for higher visibility, but they inevitably lead to a low signal-to-noise, hence requiring a potentially very large number of experimental runs for acquiring statistically significant data. Large occupations, on the other hand, lead to higher signal-to-noise, but also to a degradation of the visibility towards the nonclassical threshold of $V = 0.5$. The mode population of ~ 0.1 resulting from our numerical simulations appears to be a reasonable compromise; following the scaling of the visibility with $\langle \hat{n} \rangle$ predicted by the simple analytic model, it appears that one could safely increase the population to ~ 0.2 before a nonclassical threshold is reached to within a typical uncertainty of ~ 13 % (as per quoted value of $V \simeq 0.69 \pm 0.09$) obtained through our simulations.

The proposal is also robust to other experimental considerations such as the implementation of the Bragg pulses; e.g., one may use square Bragg pulses rather than Gaussians. Furthermore, experimental control of the Bragg pulses is sufficiently accurate to avoid any degradation of the dip visibility. Modifying the relative timing of the π and $\pi/2$ pulses by few percent in our simulations does not explicitly affect the dip visibility, rather only the period of the oscillations in the wings of $\overline{g}_{RL}^{(2)}(t_4)$. This may lead to a systematic change in the calculated dip visibility, however, this is overwhelmed by the uncertainty of 13 % which accounts for the fluctuations of $\overline{g}_{RL}^{(2)}(t_4)$ about the mean.

Importantly, we expect that the fundamentally new aspects of the matter-wave setup, namely the multimode nature of the scattering halo and the differences from the archetypal HOM input state of $|1, 1\rangle$, as well as the specific measurement protocol we have proposed for dealing with these new aspects, are broadly applicable to other related matter-wave setups that generate pair-correlated atoms. These include molecular dissociation [19], an elongated BEC in a parametrically shaken trap [14], or degenerate four-wave mixing in an optical lattice [28, 29]. In the present work, we focus on condensate collisions only due to the accurate characterisation, both experimental and theoretical, of the atom-atom correlations, including in a variety of collision geometries [7–11].

2.4 Conclusion

In summary, we have shown that an atom-optics analogue of the Hong–Ou–Mandel effect can be realised using colliding condensates and a sequence of Bragg pulses. The HOM dip visibility greater than 50 % implies that the atom-atom correlations in this process cannot be described by classical stochastic random variables. Generation and detection of such quantum correlations in matter waves can serve as precursors to stronger tests of quantum mechanics such as those implied by a Bell inequality violation and the Einstein–Podolsky–Rosen paradox [30]. In particular, the experimental demonstration of the atom-optics HOM effect would serve as a suitable starting point to experimentally demonstrate a violation of a Bell inequality using an atom-optics adaptation of the Rarity–Tapster setup [17]. In this setup, one would tune the Bragg pulses as to realise two separate HOM-interferometer arms, enabling to mix *two* angle-resolved pairs of momentum modes from the collisional halo, such as $(\mathbf{k}, \mathbf{q})$ and $(-\mathbf{k}, -\mathbf{q})$, which would then form the basis of a Bell state $|\Psi\rangle = \frac{1}{\sqrt{2}}(|\mathbf{k}, -\mathbf{k}\rangle + |\mathbf{q}, -\mathbf{q}\rangle)$ [31].

References

1. Hong, C.K., Ou, Z.Y., Mandel, L.: Measurement of subpicosecond time intervals between two photons by interference. Phys. Rev. Lett. **59**, 2044–2046 (1987)
2. Ou, Z.Y., Mandel, L.: Violation of Bell's inequality and classical probability in a two-photon correlation experiment. Phys. Rev. Lett. **61**, 50–53 (1988)
3. Knill, E., Laflamme, R., Milburn, G.J.: A scheme for efficient quantum computation with linear optics. Nature **409**, 46 (2001)

4. Kok, P., Lee, H., Dowling, J.P.: Creation of large-photon-number path entanglement conditioned on photodetection. Phys. Rev. A **65**, 052104 (2002)
5. Duan, L.-M., Monroe, C.: Colloquium: quantum networks with trapped ions. Rev. Mod. Phys. **82**, 1209–1224 (2010)
6. Lang, C., et al.: Correlations, indistinguishability and entanglement in Hong-Ou-Mandel experiments at microwave frequencies. Nat. Phys. **9**, 345–348 (2013)
7. Perrin, A., et al.: Observation of atom pairs in spontaneous four-wave mixing of two colliding Bose-Einstein condensates. Phys. Rev. Lett. **99**, 150405 (2007)
8. Perrin, A., et al.: Atomic four-wave mixing via condensate collisions. New J. Phys. **10**, 045021 (2008)
9. Krachmalnicoff, V., et al.: Spontaneous four-wave mixing of de Broglie waves: beyond optics. Phys. Rev. Lett. **104**, 150402 (2010)
10. Jaskula, J.-C., et al.: Sub-Poissonian number differences in four-wave mixing of matter waves. Phys. Rev. Lett. **105**, 190402 (2010)
11. Kheruntsyan, K.V., et al.: Violation of the Cauchy-Schwarz inequality with matter waves. Phys. Rev. Lett. **108**, 260401 (2012)
12. Kozuma, M., et al.: Coherent splitting of Bose-Einstein condensed atoms with optically induced Bragg diffraction. Phys. Rev. Lett. **82**, 871–875 (1999)
13. Meystre, P.: Atom Optics. Springer, New York (2001)
14. Bücker, R., et al.: Twin-atom beams. Nat. Phys. **7**, 608 (2011)
15. Lücke, B., et al.: Twin matter waves for interferometry beyond the classical limit. Science **334**, 773 (2011)
16. Gross, C., et al.: Atomic homodyne detection of continuous-variable entangled twin-atom states. Nature **480**, 219 (2011)
17. Rarity, J.G., Tapster, P.R.: Experimental violation of Bell's inequality based on phase and momentum. Phys. Rev. Lett. **64**, 2495–2498 (1990)
18. Norrie, A.A., Ballagh, R.J., Gardiner, C.W.: Quantum turbulence and correlations in Bose-Einstein condensate collisions. Phys. Rev. A **73**, 043617 (2006)
19. Savage, C.M., Schwenn, P.E., Kheruntsyan, K.V.: First-principles quantum simulations of dissociation of molecular condensates: atom correlations in momentum space. Phys. Rev. A **74**, 033620 (2006)
20. Deuar, P., Drummond, P.D.: Correlations in a BEC collision: first-principles quantum dynamics with 150 000 atoms. Phys. Rev. Lett. **98**, 120402 (2007)
21. Mølmer, K., et al.: Hanbury Brown and Twiss correlations in atoms scattered from colliding condensates. Phys. Rev. A **77**, 033601 (2008)
22. Ogren, M., Kheruntsyan, K.V.: Atom-atom correlations in colliding Bose-Einstein condensates. Phys. Rev. A. **79**, 021606 (2009)
23. Deuar, P., Chwedeńczuk, J., Trippenbach, M., Ziń, P.: Bogoliubov dynamics of condensate collisions using the positive-P representation. Phys. Rev. A **83**, 063625 (2011)
24. Batelaan, H.: The Kapitza-Dirac effect. Contemp. Phys. **41**, 369–381 (2000)
25. Rarity, J.G., Tapster, P.R.: Two-color photons and nonlocality in fourth-order interference. Phys. Rev. A **41**, 5139–5146 (1990)
26. Braunstein, S.L., van Loock, P.: Quantum information with continuous variables. Rev. Mod. Phys. **77**, 513–577 (2005)
27. Ögren, M., Kheruntsyan, K.V.: Role of spatial inhomogeneity in dissociation of trapped molecular condensates. Phys. Rev. A **82**, 013641 (2010)
28. Hilligsøe, K.M., Mølmer, K.: Phase-matched four wave mixing and quantum beam splitting of matter waves in a periodic potential. Phys. Rev. A **71**, 041602 (2005)
29. Bonneau, M., et al.: Tunable source of correlated atom beams. Phys. Rev. A **87**, 061603 (2013)
30. Kofler, J., et al.: Einstein-Podolsky-Rosen correlations from colliding Bose-Einstein condensates. Phys. Rev. A **86**, 032115 (2012)
31. Lewis-Swan, R.J., Kheruntsyan, K.V.: to be published (see also the Book of Abstracts of ICAP 2012 – The 23rd International Conference on Atomic Physics, 23–27 July 2012, Ecole Polytechnique, Palaiseau, France)

Chapter 3
Proposal for a Motional-State Bell Inequality Test with Ultracold Atoms

In Sect. 1.4.2 and Chap. 2 we outlined how, in the simplest approximation, the process of spontaneous four-wave mixing via condensate collisions produces a multi-mode analog of the two-mode squeezed vacuum state. Such a state exhibits non-classical correlations, which, when combined with an appropriate measurement scheme, can be used to demonstrate a violation of a Bell inequality. In this chapter, we propose such a demonstration by realization of an atom-optics analog of a Rarity–Tapster interferometer, which was previously used in quantum optics to demonstrate a successful violation using momentum-entangled photons [1]. Our investigation is focused on the feasibility of such a demonstration in realistic experimental regimes and responds to many of the key research questions of this thesis, such as how the violation depends on various experimental parameters and the robustness of simple 'toy-model' results. In particular, we illustrate that while the idealized results outlined in Sect. 1.4.1 for the simple two-mode squeezed vacuum are a reasonably valid estimate, understanding the influence of generic differences between the atom-optics and quantum optics schemes, such as the multi-mode collision halo and the use of Bragg pulses (the atom-optics equivalent of optical beam-splitters and mirrors), prove crucial in realizing a violation in realistic systems.

The remainder of this chapter was originally published as: *'Proposal for a motional-state Bell inequality test with ultracold atoms'* [R.J. Lewis-Swan and K.V. Kheruntsyan, Phys. Rev. A **91**, 052114 (2015)]. Supplementary material for the paper can be found in Appendix D.

3.1 Introduction

Bell inequalities [2, 3] have arguably been regarded as "the most profound discovery in science" [4]. They provide a fundamental distinction between local hidden-variable (LHV) descriptions of physical reality and the description based on quan-

© Springer International Publishing Switzerland 2016   57
R.J. Lewis-Swan, *Ultracold Atoms for Foundational Tests of Quantum Mechanics*, Springer Theses, DOI 10.1007/978-3-319-41048-7_3

tum mechanics wherein the concept of nonlocal entanglement is a fundamental ingredient. Violations of Bell inequalities, which reject all LHV theories and attest for the validity of quantum mechanics, have been demonstrated in numerous experiments with massless photons [1, 5–8], but in only a handful of experiments involving massive particles [9, 10]. In addition, all massive particle experiments have so far been restricted to exploiting entanglement between internal (spin) degrees of freedom, but never between external (motional) degrees of freedom such as translational momentum. Here, we propose and simulate a matter-wave experiment which, for the first time, can demonstrate a Bell inequality violation for pairs of momentum-entangled ultracold atoms produced in a collision [11–14] of two Bose-Einstein condensates. In such a motional-state Bell inequality test, particle masses become directly relevant, thus enabling extensions of fundamental tests of quantum mechanics into regimes which may involve couplings to gravitational fields and hence find connections to theories of gravitational decoherence [15]. This is important in view of future possible tests of quantum mechanics or its modifications (which currently go beyond established theories) in an attempt to resolve the current incompatibility of quantum mechanics and the theory of gravity.

The original Bell inequality was formulated by John Bell [2, 3] in response to Einstein, Podolsky, and Rosen's (EPR) argument [16, 17] that, under the premises of local realism, quantum mechanics appears to be incomplete and hence must be supplemented by hidden variables in order to explain the 'spooky-action-at-a-distance' due to entanglement between space-like separated particles. The first conclusive experimental demonstrations of Bell inequality violations with photons were reported in the early 1980s through to 1990s [1, 6–8] and used sources of pair-correlated photons, such as from a radiative cascade or parametric down-conversion. It took almost another two decades before the first massive-particle Bell violations emerged, utilising pairs of trapped ions [9] or proton pairs from the radiative decay of metastable ^{2}He [10]. These experiments all relied on entanglement between the internal degrees of freedom—either the photon polarizations or the particle spins, with the notable exception of the Rarity–Tapster experiment [1] which explored entanglement between photons momenta (see also [18]).

In recent years, there has been an increasing number of experiments, particularly in the field of ultracold atoms [19–21] and opto-mechanics [22], generating and quantifying various forms of massive-particle entanglement [23–25]. However, these should be distinguished from experiments designed to rule out LHV theories via a Bell inequality violation—the most stringent test of quantum mechanics. Ultracold atoms, nevertheless, provide a promising platform for extending these experiments towards Bell inequality tests [26–29], due to their high degree of isolation from the environment and the existing high degree of control over system parameters, including the internal and external degrees of freedom.

Our proposal for a motional-state Bell inequality uses pair-correlated atoms from colliding Bose-Einstein condensates (BECs) and in this respect represents an ultimate successor to recent experiments demonstrating sub-Poissonian relative atom number statistics, violation of the classical Cauchy–Schwartz inequality [13, 14], atomic Hong–Ou–Mandel effect [30, 31], and a recent theoretical proposal for demonstrating the EPR paradox [32] using the same collision process. A closely related process of dissociation of diatomic molecules has been recently proposed in Ref. [26] for demonstrating a Bell violation based on energy-time entanglement; the same process of molecular dissociation was previously discussed in Ref. [33] in the context of the EPR paradox for atomic quadrature measurements.

3.2 Proposed Atomic Rarity–Tapster Setup

The schematic diagram of the proposed experiment is shown in Fig. 3.1. A highly elongated (along the x-axis) BEC is initially split into two counterpropagating halves with momenta $\pm\mathbf{k}_0$ along z in the centre-of-mass frame [11, 12]. Constituent atoms of the condensate undergo binary elastic s-wave scattering and populate a nearly spherical scattering halo (of radius $k_r \simeq 0.95|\mathbf{k}_0|$) of pair-correlated atoms [12] via the process of spontaneous four-wave mixing. Previous experiments and theory [11–14, 34] have shown the existence of strong atom-atom correlation between pairs of diametrically opposite momentum modes, such as $(\mathbf{p}, -\mathbf{p})$ and $(\mathbf{q}, -\mathbf{q})$ (shown

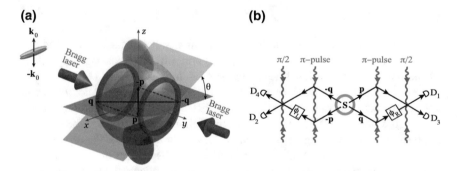

Fig. 3.1 Schematic diagram of the collision geometry and the proposed adaptation of the Rarity–Tapster scheme. **a** The two condensates in position space, counter-propagating along the z-axis with mean momenta $\pm\mathbf{k}_0$, are shown in the left, upper corner; the same condensates in momentum space (or after a time-of-flight expansion) have a pancake shape and are shown on the north and south poles of the spherical halo of scattered atoms. The counter-propagating (along y) Bragg lasers are tuned to couple and transfer the population between two pairs of momentum modes, such as the pair $(\mathbf{p}, \mathbf{q})$ and $(-\mathbf{q}, -\mathbf{p})$, indicated on the equatorial plane of the scattering halo. A similar quartet of modes (not shown for clarity), coupled by the same Bragg lasers, can be identified on any other plane obtained by rotating the equatorial plane by an angle θ around the y-axis; together, all these quartets of modes form two opposing rings shown in *red*. **b** The Rarity–Tapster scheme for implementing the π and $\pi/2$ Bragg pulses on pairs of momentum modes emanating from the source (S) and the arrangement of two independent relative phase setting ϕ_L and ϕ_R (respectively, between $\mathbf{p}$ and $\mathbf{q}$, and between $-\mathbf{p}$ and $-\mathbf{q}$) imposed in the left and the right arms of the setup. After being mixed by the final $\pi/2$ pulse, the output modes are detected by four atom detectors D_i ($i = 1, 2, 3, 4$) and different coincidence counts C_{ij} are measured for calculating the CHSH-Bell parameter S

in Fig. 3.1 on the equatorial plane of the scattering halo), similar to the correlation between twin-photons in parametric down-conversion [1, 7, 8]. After the end of the collision, we apply two separate Bragg pulses (π and $\pi/2$) tuned to couple uncorrelated atoms from each respective pair, namely $(\mathbf{p}, \mathbf{q})$ and $(-\mathbf{p}, -\mathbf{q})$. The Bragg pulses replicate the atom optics analogs of a mirror and a beam splitter (see Fig. 3.1b), thus realising the two interferometer arms of the Rarity-Tapster optical setup [1] (see also Ref. [35] which proposes the same scheme for implementing phase-sensitive measurements with ultracold atoms). A variable phase shift is additionally applied before the beam-splitter ($\pi/2$) pulse to the two lower arms of the interferometer, corresponding to a relative phase shift of ϕ_L between $-\mathbf{p}$ and $-\mathbf{q}$, and $\phi_R = \phi_L + \phi$ between $\mathbf{q}$ and $\mathbf{p}$. This replicates the poploriser angle setting or relative phase settings in the optical Bell tests of Refs. [1, 6], and can be realised by means of introducing a relative phase ϕ_L between the two counterpropagating Bragg lasers that realise the π-pulse, combined with an additional relative phase shift ϕ between the left and the right arms of the interferometer, implemented by, e.g., the well-established technique of optical phase imprinting [36, 37].

In the low-gain regime of atomic four-wave mixing (see below), this process approximately realises a prototypical Bell state of the form

$$|\Psi\rangle = \frac{1}{\sqrt{2}}(|1_{\mathbf{p}}, 1_{-\mathbf{p}}\rangle + |1_{\mathbf{q}}, 1_{-\mathbf{q}}\rangle), \tag{3.1}$$

which corresponds to a pair of atoms in a quantum superposition of belonging to either the momentum modes $\mathbf{p}$ and $-\mathbf{p}$, or $\mathbf{q}$ and $-\mathbf{q}$. By measuring appropriate second-order correlation functions using atom-atom coincidences between certain pairs of atom detectors D_i ($i = 1, 2, 3, 4$), for a chosen set of applied phases ϕ_L and ϕ_R, one can construct (see below) the CHSH-Bell parameter S for the Clauser–Horne–Shimony–Holt (CHSH) version of the Bell inequality [6, 38]. The choice of phase settings ϕ_R and ϕ_L gives rise to non-locality in the vein of the original EPR paradox as atom-atom coincidences are intrinsically dependent on both phase settings, analogous to choosing polarization directions in archetypal optics experiments [5, 6]. Indeed, the Rarity–Tapster interferometric scheme can be mapped to a spin-1/2 or polarization-entangled system [5], wherein choosing the phases ϕ_L and ϕ_R directly controls the polarization basis in which each measurement is made.

Apart from coupling two pairs of momentum modes, $(\mathbf{p}, \mathbf{q})$ and $(-\mathbf{q}, -\mathbf{p})$, shown on the equatorial plane of Fig. 3.1a, the Bragg pulses couple many other pairs of scattering modes that have the same wave-vector difference of $2\mathbf{k}_L \equiv |\mathbf{p} - \mathbf{q}| = |(-\mathbf{p}) - (-\mathbf{q})|$. Quartets of such modes, forming independent Bell states, can be identified on any other plane obtained from the equatorial plane by rotating it by an angle θ around the y-axis. Atom-atom coincidences between these modes can therefore be used as independent measurements for evaluating the respective CHSH-Bell parameter S. Averaging over many coincidence counts obtained in this way on a single scattering halo (in addition to averaging over many experimental runs) can be used to increase the signal-to-noise ratio and ultimately help the acquisition of a statistically significant result for S.

3.3 Simple Toy Model

Before presenting the results of our simulations, we make a brief diversion to discuss an important difference between the ideal prototype Bell state of the form of Eq. (3.1) and that which corresponds to the output of the simplest model of four-mode optical parametric down-conversion, to which our system can be reduced to in its most rudimentary approximation (see Refs. [34, 39, 40] and Appendix D). The Hamiltonian describing this process [41, 42] can be written as $\hat{H} = \hbar g(\hat{a}_1^\dagger \hat{a}_2^\dagger + \hat{a}_3^\dagger \hat{a}_4^\dagger + h.c.)$, where $g > 0$ is a gain coefficient, related in our context to the density ρ_0 of the initial source condensate (assumed uniform) and the s-wave interaction strength $U = 4\pi\hbar^2 a/m$ through $g = U\rho_0/\hbar$ [34, 39], where a is the s-wave scattering length. The output state of this model (for an initial vacuum state for all four modes) in the Schrödinger picture can be written in terms of an expansion in the Fock-state basis as [30, 43]

$$|\Psi\rangle = (1 - \alpha^2) \sum_{k,m=0}^{\infty} \alpha^{(k+m)} |k\rangle_1 |k\rangle_2 |m\rangle_3 |m\rangle_4, \tag{3.2}$$

where $\alpha = \tanh(gt)$ and t is the collision duration. In the weak-gain regime, which corresponds to $\alpha \simeq gt$ and hence an average mode occupation in each of the four modes ($\langle \hat{a}_i^\dagger \hat{a}_i \rangle \equiv n = \sinh^2(gt)$, $i = 1, 2, 3, 4$) of $n \simeq \alpha^2 = (gt)^2 \ll 1$, the sum over Fock states can be truncated to lowest order in α to

$$|\Psi\rangle \propto |0\rangle_1 |0\rangle_2 |0\rangle_3 |0\rangle_4$$
$$+ \alpha(|1\rangle_1 |1\rangle_2 |0\rangle_3 |0\rangle_4 + |0\rangle_1 |0\rangle_2 |1\rangle_3 |1\rangle_4). \tag{3.3}$$

Taking into account the fact that the contribution from the pure vacuum state (the first term) does not affect the outcome of any correlation (coincidence) measurements (except for reducing the absolute data acquisition rate through multiple experimental realizations), we can further approximate this state by $|\Psi\rangle \propto \alpha(|1\rangle_1 |1\rangle_2 |0\rangle_3 |0\rangle_4 + |0\rangle_1 |0\rangle_2 |1\rangle_3 |1\rangle_4)$. Equation (3.1) corresponds to this state in a shorthand notation. Such a state can itself be mapped to the archetypical Bell state $|\Psi^+\rangle = \frac{1}{\sqrt{2}}(|+\rangle_L |-\rangle_R + |-\rangle_L |+\rangle_R)$ in the polarisation or spin-1/2 $\hat{S}_z$ basis, where the subscript (L, R) refers to the left and right arms of the interferometer and $+$ $(-)$ refer to the upper (lower) paths, in terms of the diagram of Fig. 3.1b of the main text.

This ideal Bell state gives a maximal value of $S = 2\sqrt{2}$ (for a definition of the CHSH-Bell parameter S, see Sect. 3.4) and hence a maximal Bell violation ($S > 2$) by definition. However, in general, when using spontaneous parametric down-conversion as a suitable source of pair correlated particles, one must keep in mind the contribution from the higher-order Fock states (whose relative weight is very small for $n \ll 1$, implying that the contribution of events that produce, e.g., two or more photons in each of the correlated modes is extremely unlikely), leading to a breakdown of the mapping of the full state Eq. (3.2) to Eq. (3.1) and thus a reduction in S from the maximum value of $2\sqrt{2}$ to

$$S = 2\sqrt{2}\frac{1+n}{1+3n}. \tag{3.4}$$

This expression corresponds, in fact, to the full output state, Eq. (3.2), without any truncation of higher-order Fock states, and hence is valid for arbitrary n; it follows (see Appendix D) from the maximally valued anomalous moment $|m|^2 \equiv |\langle \hat{a}_1 \hat{a}_2 \rangle|^2 = |\langle \hat{a}_3 \hat{a}_4 \rangle|^2 = n(n+1)$, which is the case for this simple parametric down-conversion model [39, 40], where $n = \sinh^2(gt)$.

Equation (3.4) is an insightful result from the simplest analytic treatment as it shows the scaling of S with the mode population: for $n \ll 1$ we indeed obtain a nearly maximal Bell violation, $S \simeq 2\sqrt{2}$ whilst we find an upper bound of $n = n_{cr} = (\sqrt{2}-1)/(3-\sqrt{2}) \simeq 0.26$ beyond which the violation is no longer observed as $S \le 2$ for $n \ge n_{cr}$. We thus conclude that, for a large Bell violation, it is necessary to work in the low-gain, low mode-occupation regime of $n \ll 1$, which has, however, a practical inconvenience of requiring a large number of repeated experimental runs for achieving statistically significant data acquisition rate.

3.4 Stochastic Bogoliubov Simulations: Results and Discussion

To simulate the generation and detection of Bell states via the proposed scheme we use the stochastic Bogoliubov approach in the positive P-representation [12, 44], in which the scattered atoms are described by a small fluctuating component $\hat{\delta}(\mathbf{r}, t)$ in the expansion of the full field operator $\hat{\Psi}(\mathbf{r}, t) = \psi_0(\mathbf{r}, t) + \hat{\delta}(\mathbf{r}, t)$, where $\psi_0(\mathbf{r}, t)$ is the mean-field component describing the source condensate assumed to be in a coherent state of total average number N, initially in the ground state of the confining trap potential. This approach has previously been used to accurately model a number of condensate collision experiments, including the measurement and characterisation of atom-atom correlations via sub-Poissonian relative number statistics [13], violation of the classical Cauchy–Schwarz inequality [14], and more recently in a theoretical proposal for demonstrating an atomic Hong–Ou–Mandel effect [30]. The positive P-representation has also been used in Ref. [45] for direct probabilistic sampling of an idealised, polarisation-entangled Bell state to show how a Bell inequality violation can be simulated using the respective phase-space distribution function. Complementary to Ref. [45], we do not assume any pre-existing Bell state in our analysis, but adopt an operational approach of calculating a set of pair-correlation functions C_{ij} that define the CHSH-Bell parameter S, after real-time simulations of the collision dynamics and the application of Bragg pulses. (For the most recent formulation of the stochastic positive-P equations that we simulate, including the application of the lattice potential imposed by the Bragg lasers, see the Methods section of Ref. [30].)

The CHSH-Bell parameter S corresponding to our measurement protocol, performed for four pairs of phase settings, is defined as [1, 38]

$$S = |E(\phi_L, \phi_R) - E(\phi_L, \phi_R') + E(\phi_L', \phi_R) + E(\phi_L', \phi_R')|, \quad (3.5)$$

where

$$E(\phi_L, \phi_R) \equiv \left. \frac{C_{14} + C_{23} - C_{12} - C_{34}}{C_{14} + C_{23} + C_{12} + C_{34}} \right|_{\phi_L, \phi_R}. \quad (3.6)$$

Here, the correlation functions C_{ij} are given by $C_{ij} = \langle \hat{N}_i \hat{N}_j \rangle$, where the operator $\hat{N}_i(t) = \int_{\mathcal{V}(\mathbf{k}_i)} d^3\mathbf{k}\, \hat{n}(\mathbf{k}, t)$ corresponds to the number of atoms detected in a detection bin with dimensions Δk_d ($d = x, y, z$) and volume $\mathcal{V}(\mathbf{k}_i) = \prod_d \Delta k_d$, centered around the targeted momenta $\mathbf{k}_i$ ($i = 1, 2, 3, 4$); the set of momenta $\{\mathbf{k}_1, \mathbf{k}_2, \mathbf{k}_3, \mathbf{k}_4\}$ correspond, respectively, to $\{\mathbf{p}, -\mathbf{p}, \mathbf{q}, -\mathbf{q}\}$ used in the diagram of Fig. 3.1, while $\hat{n}(\mathbf{k}, t) = \hat{a}^\dagger(\mathbf{k}, t)\hat{a}(\mathbf{k}, t)$ is the momentum-space density, with $\hat{a}(\mathbf{k}, t)$ being the Fourier component of the field operator $\hat{\delta}(\mathbf{r}, t)$ describing the scattered atoms. The CHSH-Bell inequality states that any LHV theory satisfies an upper bound given by $S \leq 2$, irrespective of the phase settings ϕ_L, ϕ_R, ϕ_L', and ϕ_R'.

The results of our numerical simulations of the collision dynamics and ensuing Bragg pulses are shown in Figs. 3.2 and 3.3. Figure 3.2 illustrates the momentum

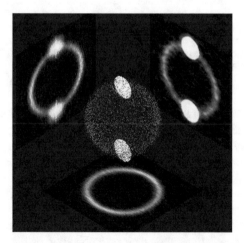

Fig. 3.2 Illustration of typical results for the collisional halo in momentum space from the stochastic Bogoliubov approach in the positive-P representation. Shown here are three orthogonal slices (cuts through the origin) of the 3D momentum distribution $n(\mathbf{k})$ at the end of the collision; the saturated (*white*) regions of the colour map correspond to the high-density colliding condensates. The central figure is a discretised scatter plot of the 3D data (shown only for illustrational purposes and comparison with Fig. 3.1), in which the dots (pixels) represent random samples of the average, but still fluctuating within the sampling error, density distribution binned into pixels whose colour coding scales with the atom number in the bin (only four color grades were used for clarity). For quantitative details of the same data on the equatorial plane, see Fig. 3.3

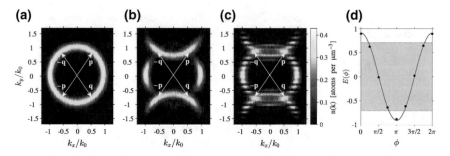

Fig. 3.3 Momentum distribution $n(\mathbf{k})$ of scattered atoms on the equatorial plane of the halo and the correlation coefficient E. Panel **a** shows the momentum distribution after the collision, at $t_1 = 65\,\mu s$; **b**—after the π-pulse chosen here to be a Gaussian, centred at $t_2 = 79\,\mu s$ and having a duration (rms width) of $\tau_\pi = 3.5\,\mu s$; and **c**—after the final $\pi/2$ pulse, centered at $t_3 = 139\,\mu s$ and having a duration of $\tau_{\pi/2} = 3.5\,\mu s$. The momentum axes $k_{x,y}$ are normalised to the collision momentum $k_0 \equiv |\mathbf{k_0}|$ (in wave-number units), which in our simulations is $k_0 = 4.7 \times 10^6\,\mathrm{m}^{-1}$. The plotted results are for an initial BEC containing a total average number of $N = 1.9 \times 10^4$ atoms of metastable helium ($^4\mathrm{He}^*$) prepared in a harmonic trap of frequencies $(\omega_x, \omega_y, \omega_z)/2\pi = (64, 1150, 1150)$ Hz and colliding with the scattering length of $a = 5.3$ nm; all these parameters are very close to those realised in recent experiments [12–14]. The optimal timing of the final Bragg pulse differs slightly for condensates with different N; in particular, t_3 ranged from 135.5 to 139 μs for the data in Fig. 3.4 (see Appendix D). The data is averaged over $\sim$30,000 stochastic trajectories on a spatial lattice of $722 \times 192 \times 168$ points. Panel **d** shows the correlation coefficient $E(\phi_L, \phi_R)$ as a function of $\phi \equiv \phi_L - \phi_R$, for the same detection bin sizes as in Fig. 3.4, *blue circles*. The data points are from numerical simulations (error bars of two standard deviations, representing sampling errors from 360 stochastic runs, are within the marker size), including averaging over $\sim$370 quartets of distinct detection volumes on the two opposing rings of the scattering halo shown in Fig. 3.1, while the solid line is from the Gaussian-fit model, Eq. (3.7). A maximum amplitude of $E_0 > 1/\sqrt{2}$ (outside the *shaded region*) corresponds to a correlation strength that can lead to a Bell inequality violation, given the underlying sinusoidal behaviour

space density distribution of the collisional halo, while Fig. 3.3 focuses on the quantitative results on the equatorial plane, for: (a) at the end of the collision; (b) after the application of the *pi* pulse, and (c) after the $\pi/2$ puls. The upper and lower semicircles in (b) correspond to Bragg-kicked populations between the targeted momenta around **p** and **q**, and between $-\mathbf{q}$ and $-\mathbf{p}$, while (c) shows the final distribution after mixing. The density modulation in (c) (in parts of the halo lying outside the vicinity of the targeted momentum modes, where the transfer of population during the π pulse is not 100% efficient) is simply the result of interference between the residual and transferred atomic populations upon their recombination on the beamsplitter [30].

We next use the stochastic Bogoliubov simulations to calculate the atom-atom correlations C_{ij}, for the optimal choice of phase angles $\phi_L = 0$, $\phi'_L = \pi/2$, $\phi_R = \pi/4$, and $\phi'_R = 3\pi/4$ [1]. The dependence of the resulting correlation coefficient E on the relative phase $\phi \equiv \phi_L - \phi_R$ is shown in Fig. 3.3d; it displays a sinusoidal dependence $E_0 \cos \phi$ which can also be predicted from a simple Gaussian-fit analytic model (see Appendix D):

$$E(\phi_L, \phi_R) = \frac{h \prod_d \alpha_d}{h \prod_d \alpha_d + 2 \prod_d (\lambda_d)^2} \cos(\phi_L - \phi_R). \tag{3.7}$$

In this model, C_{ij} is expressed in terms of the density-density correlation function $G^{(2)}(\mathbf{k}, \mathbf{k}', t_1) = \langle \hat{a}^\dagger(\mathbf{k}, t_1) \hat{a}^\dagger(\mathbf{k}', t_1) \hat{a}(\mathbf{k}', t_1) \hat{a}(\mathbf{k}, t_1) \rangle$ after the collision as $C_{ij} = \int_{V(\mathbf{k}_i)} d^3\mathbf{k} \int_{V(\mathbf{k}_j)} d^3\mathbf{k}' G^{(2)}(\mathbf{k}, \mathbf{k}', t_1)$, and we use the fact that $G^{(2)}(\mathbf{k}, \mathbf{k}', t_1)$ itself is typically well approximated [11, 14, 46] by a Gaussian function of the form $G^{(2)}(\mathbf{k}, \mathbf{k}', t_1) = \bar{n}^2 (1 + h \prod_d \exp[-(k_d + k'_d)^2 / 2\sigma_d^2])$, where we have assumed that the density of scattered atoms is approximately constant over the integration volume and is given by $\bar{n}$. Thus, in Eq. (3.7), h is the height (above the background level of $\bar{n}^2$) of the pair correlation $G^{(2)}(\mathbf{k}, \mathbf{k}', t_1)$, σ_d is the rms width, $\lambda_d \equiv \Delta k_d / 2\sigma_d$ is the relative bin size, and $\alpha_d \equiv (e^{-2\lambda_d^2} - 1) + \sqrt{2\pi}\lambda_d \, \mathrm{erf}\left(\sqrt{2}\lambda_d\right)$. The particular form of E in Eq. (3.7) is obtained from this model by assuming the subsequent 'mirror' and 'beam-splitter' mix the coupled modes exactly. The visibility of the correlation coefficient E bounds the maximum attainable violation of the CHSH-Bell inequality for a specific set of phase settings, with a lower-limit of $E_0 = 1/\sqrt{2}$ required for $S > 2$, and a maximum value of $E_0 = 1$ corresponding to $S = 2\sqrt{2}$.

The results of calculations of the CHSH-Bell parameter S are shown in Fig. 3.4, where we explore its dependence on the strength of atom-atom correlations and the detection bin size. The dependence on the correlation strength, for a fixed collision velocity and trap frequencies, reflects essentially the dependence on the peak density of the initial BEC, which itself depends on the total average number of atoms loaded in the trap [34]. The results of stochastic simulations in Fig. 3.4b are plotted alongside the predictions of the Gaussian-fit analytic model, which from Eq. (3.7) gives

$$S = 2\sqrt{2} \frac{h \prod_d \alpha_d}{h \prod_d \alpha_d + 2 \prod_d (\lambda_d)^2}. \tag{3.8}$$

As we see, the analytic prediction agrees reasonably well with the numerical results, both showing that strong Bell violations are favoured for: (*i*) smaller condensates, leading to lower mode population in the scattering halo and thus higher correlation strength, and (*ii*) smaller bin sizes, for which the strength of atom number correlations does not get diluted due to the finite detection resolution. The discrepancies between the numerical and analytic results are due the fact that the analytic model assumes uniform halo density across the integration bin and perfect Bragg pulses, both in terms of the intended transfer efficiency and its insensitivity to the momentum offsets within the integration bin, whereas the numerical simulations are performed with realistic Bragg pulses acting on the actual inhomogeneous scattering halo. Nevertheless, an important conclusion that we reach here is that the Bell violation in our scheme can tolerate experimentally relevant imperfections that are often ignored in oversimplified models.

The general form of Eq. (3.8) displays similar behaviour to that obtained in the simple model of four-mode parametric down-conversion, Eq. (3.4). As previously, it gives a simple and insightful picture in terms of the dependence of the expected value

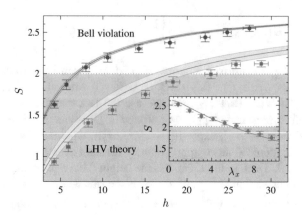

Fig. 3.4 CHSH-Bell parameter S as a function of the correlation strength h (see text); the value of h can be controlled by varying the total average atom number N in the initial BEC. For the data points shown here, N was varied between 1.9×10^4 (largest h) and 7.4×10^4 (smallest h). The two sets of data correspond to two different detection bin sizes: $(\Delta k_x, \Delta k_y, \Delta k_z) = (0.052, 0.53, 0.47)\ \mu m^{-1}$—*blue circles*, and $(0.12, 1.24, 1.10)\ \mu m^{-1}$—*red squares*. The vertical error bars on data points indicate the stochastic sampling errors; the horizontal error bars are the sampling errors on the value of h. Each individual data point is a result of averaging over approximately 2000 stochastic trajectories simulated on a computational lattice of $722 \times 192 \times 168$ points, which were run on Intel E5-2660 Xeon CPUs taking a total of $\sim$15 hours on a 128-core cluster, or $\sim$2000 CPU hours. The results are compared to the analytic predictions (*solid lines*) of Eq. (3.8); uncertainty (*shaded regions*) is due to the uncertainty in determining σ_d. The *inset* shows the explicit dependence of S on Δk_x (in units of $2\sigma_x = 0.068\ \mu m^{-1}$), for fixed $(\Delta k_y, \Delta k_z) = (0.77\sigma_y, 0.89\sigma_z) = (0.53, 0.47)\ \mu m^{-1}$ and $N = 1.9 \times 10^4$ ($h \simeq 27$). For a typical time-of-flight expansion time of $t_{exp} \sim$300 ms, which maps the atomic momentum distribution into position space density distribution, and which is when the atoms are experimentally detected, these detection bin sizes convert to position space distances of $(\Delta x, \Delta y, \Delta z) \simeq (0.32, 2.5, 2.2)$ mm (where we have taken $\lambda_x = 1$ for definitiveness), which are several times larger than the three orthogonal resolutions of multichannel plate detectors used in ^{4}He* experiments [14, 47]

of S on just a few parameters at the end of the collision—the correlation widths, the correlation height and the detection bin size. As we see from the comparison of the predictions of Eq. (3.8) to the actual numerical results in Fig. 3.4, the agreement is remarkable for such a simple analytic result. The scaling with the halo mode occupation, as that in Eq. (3.4), is no longer explicit, but it now emerges most simply through the detection bin size, wherein a smaller bin size gives a smaller average number of detected atoms and hence larger values of S as seen in the inset of Fig. 3.4. Similarly, such a scaling emerges through the height of the correlation h: the correlation is typically stronger for four-wave mixing regimes that produce collisional halo of smaller density or smaller bin occupation (for a fixed bin size), leading to larger values of S. In the four-mode down-conversion model, where the relevant normalized pair-correlation function is given by $g_{12}^{(2)} = g_{34}^{(2)} = 2 + 1/n$ [39, 40] and therefore $h = 1 + 1/n$, this corresponds to $h \gg 1$ which is again the regime of low mode occupation $n \ll 1$ as we discussed previously.

We further emphasise that the general applicability of our Gaussian-fit analytic model and, in particular, the relatively simple result of Eq. (3.8) are not limited to condensate collision experiments. Rather, these results can be applied to any other ultracold atom experiment—a candidate for a Bell test—as long is it produces two pair-correlated 'scattering' modes that can be approximated by Gaussian correlation functions and subsequently subjected to 'mirror' and 'beam-splitter' pulses as to realise an atomic Tapster interferometer.

3.5 Conclusions

In summary, we have shown that condensate collisions are a promising platform for testing motional-state Bell inequalities with massive particles. We predict a CHSH-Bell inequality violation ($S > 2$) for a range of parameters well within currently accessible experimental regimes.

Our numerical simulations take into account a range of physically important processes beyond the common analysis of oversimplified toy models. Importantly this includes: (i) the multimode nature of the colliding Bose-Einstein condensates and subsequent scattering halo; (ii) the spatial expansion and separation of the source condensates during the collision and hence during the pair production process (for comparison, the 'pump mode' in the optical down-conversion case remains practically unchanged in the required weak-gain regime); and (iii) the fact that the atomic 'mirror' and 'beam-splitter' Bragg pulses act, in fact, as momentum kicks (translations) rather than as actual (optical) reflections. By modelling the real-time application of the Bragg pulses, without assuming ideal π- and $\pi/2$-pulses (100 % and 50 % transfer respectively), we implicitly allow for small amounts of losses (hence decoherence) into higher-order Bragg scattering modes. We also take into account the nontrivial effects of phase dispersion, absent in photonic experiments, by optimising the timing and application of the Bragg pulses in the interferometer. Remarkably, many of these effects can also be captured via the semi-analytic Gaussian-fit model of Eqs. (3.7) and (3.8), which is found to be both qualitatively and quantitatively rather accurate.

Such detailed quantitative analysis is important for a theoretical proposal to be relevant to possible experimental demonstrations of a Bell inequality violation. This is further supported by our analysis in terms of finite detector resolution and the utilization of multiple quartets of modes in our calculations: increasing the rate of data acquisition is crucial for experiments with ultracold atoms which typically have relatively slow duty cycles of the order of half a minute (for comparison, the repetition rates of a pump laser in modern optical parametric down-conversion experiments can reach tens of MHz).

A laboratory demonstration of such a violation would be a major advance in experimental quantum physics as it would lead to a better understanding of massive particle entanglement involving motional states. Apart from extending foundational tests of quantum mechanics into new regimes, such experiments can potentially lead to an opening of a new experimental agenda, such as testing the theories of decoherence due to coupling to gravitational fields [15] and answering questions that are relevant to the understanding of the interplay between quantum theory and gravity and their possible unification.

References

1. Rarity, J.G., Tapster, P.R.: Experimental violation of Bell's inequality based on phase and momentum. Phys. Rev. Lett. **64**, 2495–2498 (1990)
2. Bell, J.S.: On the Einstein–Podolsky–Rosen paradox. Phys. (N.Y.) **1**, 195 (1964)
3. Bell, J.S.: Speakable and Unspeakable in Quantum Mechanics, Cambridge University Press, Cambridge (1987)
4. Stapp, H.P.: Bell's theorem and world process. Nuovo Cimento B **29**, 270–276 (1975)
5. Aspect, A., Grangier, P., Roger, G.: Experimental realization of Einstein–Podolsky–Rosen–Bohm gedankenexperiment: a new violation of Bell's inequalities. Phys. Rev. Lett. **49**, 91–94 (1982)
6. Aspect, A., Dalibard, J., Roger, G.: Experimental test of Bell's inequalities using time-varying analyzers. Phys. Rev. Lett. **49**, 1804–1807 (1982)
7. Ou, Z.Y., Mandel, L.: Violation of Bell's inequality and classical probability in a two-photon correlation experiment. Phys. Rev. Lett. **61**, 50–53 (1988)
8. Weihs, G., Jennewein, T., Simon, C., Weinfurter, H., Zeilinger, A.: Violation of Bell's inequality under strict Einstein locality conditions. Phys. Rev. Lett. **81**, 5039–5043 (1998)
9. Rowe, M.A., et al.: Experimental violation of a Bell's inequality with efficient detection. Nature **409**, 791–794 (2001)
10. Sakai, H., et al.: Spin correlations of strongly interacting massive fermion pairs as a test of Bell's inequality. Phys. Rev. Lett. **97**, 150405 (2006)
11. Perrin, A., et al.: Observation of atom pairs in spontaneous four-wave mixing of two colliding Bose–Einstein condensates. Phys. Rev. Lett. **99**, 150405 (2007)
12. Krachmalnicoff, V., et al.: Spontaneous four-wave mixing of de Broglie waves: beyond optics. Phys. Rev. Lett. **104**, 150402 (2010)
13. Jaskula, J.-C., et al.: Sub-Poissonian number differences in four-wave mixing of matter waves. Phys. Rev. Lett. **105**, 190402 (2010)
14. Kheruntsyan, K.V., et al.: Violation of the Cauchy–Schwarz inequality with matter waves. Phys. Rev. Lett. **108**, 260401 (2012)
15. Penrose, R.: On gravity's role in quantum state reduction. Gen. Relativ. Gravit. **28**, 581–600 (1996)
16. Einstein, A., Podolsky, B., Rosen, N.: Can quantum-mechanical description of physical reality be considered complete? Phys. Rev. **47**, 777–780 (1935)
17. Bohm, D.: A suggested interpretation of the quantum theory in terms of hidden variables. I. Phys. Rev. **85**, 166–179 (1952)
18. Howell, J.C., Bennink, R.S., Bentley, S.J., Boyd, R.W.: Realization of the Einstein–Podolsky–Rosen paradox using momentum- and position-entangled photons from spontaneous parametric down conversion. Phys. Rev. Lett. **92**, 210403 (2004)
19. Estève, J., Gross, C., Weller, A., Giovanazzi, S., Oberthaler, M.K.: Squeezing and entanglement in a BoseEinstein condensate. Nature **455**, 1216–1219 (2008)
20. Riedel, M.F., et al.: Atom-chip-based generation of entanglement for quantum metrology. Nature **464**, 1170 (2010)
21. Lücke, B., et al.: Twin matter waves for interferometry beyond the classical limit. Science **334**, 773 (2011)
22. Lee, K.C., et al.: Entangling macroscopic diamonds at room temperature. Science **334**, 1253 (2011)

23. Julsgaard, B., Kozhekin, A., Polzik, E.S.: Experimental long-lived entanglement of two macroscopic objects. Nature **413**, 400 (2001)
24. Matsukevich, D.N., et al.: Entanglement of remote atomic qubits. Phys. Rev. Lett. **96**, 030405 (2006)
25. Monz, T., et al.: 14-qubit entanglement: creation and coherence. Phys. Rev. Lett. **106**, 130506 (2011)
26. Gneiting, C., Hornberger, K.: Bell test for the free motion of material particles. Phys. Rev. Lett. **101**, 260503 (2008)
27. Mullin, W.J., Laloë, F.: Interference of bose-einstein condensates: quantum nonlocal effects. Phys. Rev. A **78**, 061605 (2008)
28. Laloë, F., Mullin, W.J.: Interferometry with independent Bose–Einstein condensates: parity as an EPR/Bell quantum variable. Eur. Phys. J. B **70**, 377–396 (2009)
29. Pelisson, S., Pezzé, L. & Smerzi, A. Nonlocality with ultracold atoms in a lattice. arXiv preprint arXiv:1505.02902 (2015)
30. Lewis-Swan, R.J., Kheruntsyan, K.V.: Proposal for demonstrating the Hong-Ou-Mandel effect with matter waves. Nat. Commun. **5**, 3752 (2014)
31. Lopes, R., et al.: Atomic Hong-Ou-Mandel experiment. Nature **520**, 66 (2015)
32. Kofler, J., et al.: Einstein–Podolsky–Rosen correlations from colliding Bose–Einstein condensates. Phys. Rev. A **86**, 032115 (2012)
33. Kheruntsyan, K.V., Olsen, M.K., Drummond, P.D.: Einstein–Podolsky–Rosen correlations via dissociation of a molecular Bose–Einstein condensate. Phys. Rev. Lett. **95**, 150405 (2005)
34. Perrin, A., et al.: Atomic four-wave mixing via condensate collisions. New J. Phys. **10**, 045021 (2008)
35. Kitagawa, T., Aspect, A., Greiner, M., Demler, E.: Phase-sensitive measurements of order parameters for ultracold atoms through two-particle interferometry. Phys. Rev. Lett. **106**, 115302 (2011)
36. Burger, S., et al.: Dark solitons in Bose–Einstein condensates. Phys. Rev. Lett. **83**, 5198–5201 (1999)
37. Denschlag, J., et al.: Generating solitons by phase engineering of a Bose–Einstein condensate. Science **287**, 97–101 (2000)
38. Clauser, J.F., Horne, M.A., Shimony, A., Holt, R.A.: Proposed experiment to test local hidden-variable theories. Phys. Rev. Lett. **23**, 880–884 (1969)
39. Ogren, M., Kheruntsyan, K.V.: Atom-atom correlations in colliding Bose–Einstein condensates. Phys. Rev. A. **79**, 021606 (2009)
40. Savage, C.M., Schwenn, P.E., Kheruntsyan, K.V.: First-principles quantum simulations of dissociation of molecular condensates: atom correlations in momentum space. Phys. Rev. A **74**, 033620 (2006)
41. Walls, D.F., Milburn, G.J.: Quantum Optics, 2nd edn. Springer, Berlin (2008)
42. Reid, M.D., Walls, D.F.: Violations of classical inequalities in quantum optics. Phys. Rev. A **34**, 1260–1276 (1986)
43. Braunstein, S.L., van Loock, P.: Quantum information with continuous variables. Rev. Mod. Phys. **77**, 513–577 (2005)
44. Deuar, P., Chwedeńczuk, J., Trippenbach, M., Ziń, P.: Bogoliubov dynamics of condensate collisions using the positive-P representation. Phys. Rev. A **83**, 063625 (2011)
45. Rosales-Zárate, L., Opanchuk, B., Drummond, P.D., Reid, M.D.: Probabilistic quantum phase-space simulation of Bell violations and their dynamical evolution. Phys. Rev. A **90**, 022109 (2014)
46. Chwedeńczuk, J., et al.: Pair correlations of scattered atoms from two colliding Bose–Einstein condensates: perturbative approach. Phys. Rev. A **78**, 053605 (2008)
47. Dall, R.G., et al.: Ideal n-body correlations with massive particles. Nat. Phys. **9**, 341 (2013)

Chapter 4
Sensitivity to Thermal Noise of Atomic Einstein–Podolsky–Rosen Entanglement

A notable demonstration of the Einstein–Podoslky–Rosen paradox was the quantum optics experiment of Ou et. al. [1]. This experiment used massless photons in a two-mode squeezed vacuum state produced by the process of spontaneous optical parametric down-conversion to demonstrate the paradox for optical quadratures, which are the closest equivalent to the original continuous variables of momentum and position used by EPR in their thought experiment. An obvious generalization of this experiment to the regime of massive particles would be via spin-changing collisions in a spinor condensate, which, in the simplest model, also produces the two-mode squeezed vacuum state. Motivated by this connection, a recent experimental demonstration of EPR entanglement in such a system was attempted by Gross et. al. [2], however, the results proved inconclusive.

In this chapter, we seek to understand whether this ambiguous result could be attributed to physically important sources of noise not present in analogous quantum optics experiments, such as a small (currently undetectable) thermal seed initially present in the $m_F = \pm 1$ substates. Specifically, our investigation focuses on how the spin-changing dynamics are altered and quantifying whether EPR entanglement can still be robustly generated. Thermal fluctuations are an important source of noise in realistic systems, and a better understanding of its effect on fundamental phenomena such as EPR entanglement is crucial to understanding the transition between between the classical world and quantum mechanics.

The remainder of this chapter is adapted from the published article: *'Sensitivity to thermal noise of atomic Einstein-Podolsky-Rosen entanglement'* [R.J. Lewis-Swan and K.V. Kheruntsyan, Phys. Rev. A **87**, 063635 (2013)]. The supplementary information of this article can be found in Appendix E.

4.1 Introduction

Entanglement has proven to be "the characteristic trait of quantum mechanics" as first coined by Schrödinger [3]. It forms the foundations of quantum information theory and quantum computing. Further, in interferometry entanglement

© Springer International Publishing Switzerland 2016

R.J. Lewis-Swan, *Ultracold Atoms for Foundational Tests of Quantum Mechanics*, Springer Theses, DOI 10.1007/978-3-319-41048-7_4

enables measurement precision to surpass the standard quantum limit [4]. This is particularly important in atom interferometry [5, 6] as atom flux is generally limited. However, the most important foundational trait of entanglement comes with its role in the Einstein-Podolsky-Rosen paradox (EPR) [7, 8]. This requires the underlying quantum correlations to be stronger than those satisfying the simpler inseparability criteria. The resulting EPR-entanglement criterion confronts the Heisenberg uncertainty relation and puts us into the context of EPR arguments that question the completeness of quantum mechanics and open the door to alternative descriptions of these correlations via local hidden variable theories [9–11]. The EPR paradox for continuous-variable quadrature observables [12] (which are analogous to the position and momentum observables originally discussed by EPR) has been demonstrated in optical parametric down-conversion [1] and most recently attempts have been made to demonstrate [2] the paradox with ensembles of massive particles generated by spin-changing collisions in a spinor Bose-Einstein condensate (BEC) [13, 14].

In this paper, we seek to provide a theoretical treatment of the recent experiment by Gross et al. [2] which reported entanglement, or quantum inseparability, of two atomic ensembles produced by spin-changing collisions in a ^{87}Rb BEC. For the BEC initially prepared in the $(F, m_F) = (2, 0)$ hyperfine state, the collisions produce correlated pairs of atoms in the $m_F = \pm 1$ sublevels. The authors observed that the resulting state was inseparable, though a measurement of a stronger EPR entanglement criterion was inconclusive. A normalized product of inferred quadrature variances of 4 ± 17 was reported, whereas a demonstration of the EPR paradox requires this quantity to be less than unity [12, 15].

The short-time dynamics of the spin-mixing process, for a vacuum initial state of the $m_F = \pm 1$ atoms, is similar to that of a spontaneous parametric down-conversion in the undepleted pump approximation. This paradigmatic nonlinear optical process is known to produce an EPR entangled twin-photon state that can seemingly violate the Heisenberg uncertainty relation for inferred optical quadratures [12]. Such a violation has previously been observed in 1992 by Ou et al. [1]. Due to the inconclusive nature of an analogous measurement of matter-wave quadratures in Ref. [2], we seek to perform a theoretical analysis of spin-changing dynamics and calculate various measures of entanglement in experimentally realistic regimes. In particular, we focus on the sensitivity of EPR entanglement to an initial population in the $m_F = \pm 1$ sublevels with thermal statistics. In the optical case this question is argued to be irrelevant as at optical frequencies and room temperatures the thermal population of the signal and idler modes is negligible, allowing us to safely approximate them as vacuum states. However, these considerations are inapplicable to ultracold atomic gases. This was highlighted recently by Melé-Messeguer et. al. [16], who quantitatively predicted the possibility of non-trivial thermal activation of the $m_F = \pm 1$ sublevels in a spin-1 BEC. Accordingly, when interpreting experimental results care must be taken in differentiating spin-mixing dynamics initiated by vacuum noise from that initiated by thermal noise or a small coherent seed [17]. To this end, our modelling of the experiment of Gross et al. [2] is more consistent with a small thermal population in the $m_F = \pm 1$ sublevels, rather than a vacuum initial state or small coherent seed. From a broader perspective, the connection between our results and the widely applicable model of parametric down-conversion highlights the generally fragile nature of atomic EPR entanglement to thermal noise, demonstrating that future experiments must be refined to overcome this problem.

4.2 The System

The experiment of Ref. [2] starts with a BEC of ^{87}Rb atoms prepared in the $(F, m_F) = (2, 0)$ state and trapped in a one-dimensional optical lattice. The lattice potential is sufficiently deep to prevent tunnelling between neighbouring wells. Furthermore, due to the relatively small number of atoms in each well, the spin-healing length is of the order of the spatial size of the condensate in the well meaning the spatial dynamics of the system are frozen, and hence we may treat the condensate in each well according to the single-mode approximation [18–20]. In this approximation the field operator $\hat{\psi}_i(\mathbf{r})$ for each component $i \equiv m_F = 0, \pm 1, \pm 2$ is expanded as $\hat{\psi}_i(\mathbf{r}) = \phi(\mathbf{r})\hat{a}_i$, where $\phi(\mathbf{r})$ is the common spatial ground state wavefunction ($\phi_i(\mathbf{r}) \equiv \phi(\mathbf{r})$) and $\hat{a}_i$ is the respective bosonic annihilation operator.

A quadratic Zeeman shift and microwave dressing of the $m_F = 0$ state is employed to energetically restrict the spin-mixing dynamics to the $m_F = 0, \pm 1$ states [2], and so for short time durations we may map the spin-2 system to an effective spin-1 Hamiltonian [21] of the form $\hat{H} = \hat{H}_{\text{inel}} + \hat{H}_{\text{el}} + \hat{H}_Z$,

$$\hat{H}_{\text{inel}} = \hbar g(\hat{a}_0^\dagger \hat{a}_0^\dagger \hat{a}_{-1} \hat{a}_1 + \hat{a}_1^\dagger \hat{a}_{-1}^\dagger \hat{a}_0 \hat{a}_0), \tag{4.1}$$

$$\hat{H}_{\text{el}} = \hbar g(\hat{n}_0 \hat{n}_1 + \hat{n}_0 \hat{n}_{-1}) \tag{4.2}$$

$$\hat{H}_Z = -\hbar p \left(\hat{n}_1 - \hat{n}_{-1} \right) - \hbar q \left(\hat{n}_1 + \hat{n}_{-1} \right) \tag{4.3}$$

where $\hat{n}_i = \hat{a}_i^\dagger \hat{a}_i$ is the particle number operator and $i = 0, \pm 1$ are referred to, respectively, as the pump and signal/idler modes from herein. We have ignored terms proportional to $\hat{N}(\hat{N} - 1)$ in $\hat{H}$ as this is a conserved quantity and contributes only a global phase rotation. The inelastic spin-changing collisions are described by $\hat{H}_{\text{inel}}$, and the remaining elastic s-wave scattering terms are grouped in $\hat{H}_{\text{el}}$, where g is the coupling constant associated with s-wave collisions [21]. For a spin-2 system, the coupling is given by $g = \frac{6}{14}(3g_4 + 4g_2) \int d^3\mathbf{r} |\phi(\mathbf{r})|^4$, where $g_F = 4\pi\hbar^2 a_F/m$ describes s-wave scattering with total spin F, characterised by scattering length a_F [21]. For comparison, for an actual spin-1 system the coupling constant would be given by $g = \frac{g_2 - g_0}{3} \int d^3\mathbf{r} |\phi(\mathbf{r})|^4$, where $g_F = 4\pi\hbar^2 a_F/m$. In our representation of $\hat{H}_{\text{el}}$ we have used the fact that the relative number difference, $\hat{n}_1 - \hat{n}_{-1}$, is a conserved quantity. The interaction with the magnetic field is described by $\hat{H}_Z$, where the linear and quadratic Zeeman effects are parametrized, respectively, by $p = g\mu_B B_0/\hbar$ and $q = p^2/\omega_{\text{HFS}}$ [22], with $\omega_{\text{HFS}}/2\pi \approx 6.835$ GHz being the hyperfine splitting frequency of ^{87}Rb [23] and g is the Landé hyperfine g-factor. For our initial conditions the relative number difference, $\hat{n}_1 - \hat{n}_{-1}$, will always be zero and hence we may ignore the linear Zeeman effect. We may also redefine the parameter q to absorb the effects of microwave level dressing (used by Gross et al. [2]) and any other fixed energy shift between the $m_F = 0$ and $m_F = \pm 1$ energy levels.

Simple analogies between the states of the signal and idler modes in spin-changing collisions and optical parametric down-conversion consider only $\hat{H}_{inel}$ in the undepleted pump approximation, however, competing mean-field ($\hat{H}_{el}$) and Zeeman ($\hat{H}_Z$) effects lead to additional dynamics [24] due to dephasing. The full Heisenberg operator equations of motion are given by

$$\frac{d\hat{a}_0}{d\tau} = -i \left[2\hat{a}_{-1}\hat{a}_1\hat{a}_0^\dagger + \left(\hat{n}_1 + \hat{n}_{-1}\right)\hat{a}_0 \right],   \qquad (4.4)$$

$$\frac{d\hat{a}_{\pm 1}}{d\tau} = -i \left[\hat{a}_0^2\hat{a}_{\mp 1}^\dagger + \left(\hat{n}_0 - q/g\right)\hat{a}_{\pm 1} \right],   \qquad (4.5)$$

where we have introduced $\tau = gt$ as dimensionless time. We see that the phase accrued in the $\hat{a}_{\pm 1}$ modes grows $\propto \left(\hat{n}_0 - q/g\right)$ whilst for the $\hat{a}_0$ mode the phase grows $\propto \left(\hat{n}_1 + \hat{n}_{-1}\right)$. In the short-time undepleted pump approximation [25], this is equivalent to a phase rotation $\hat{a}_{\pm 1} \to \hat{a}_{\pm 1}\exp[i\left(N_0 - q/g\right)\tau]$, where $N_0 = \langle\hat{n}_0(0)\rangle$ is the initial population of the $m_F = 0$ component. This rotation leads to a dynamical phase mismatch between the spinor components that decelerates the pair-production process [24]. To prevent phase mismatch in the short-time limit one can choose $q = gN_0$ in which case Eqs. (4.4) and (4.5) reduce to those of resonant down-conversion [25].

4.3   Results and Discussion

4.3.1   Population Dynamics

We first analyze the spin-changing dynamics for the case of a vacuum initial state for the signal/idler modes, and a coherent state $|\alpha_0(0)\rangle$ for the pump mode with initial number of atoms $N_0 = |\alpha_0(0)|^2$. This case can be treated in a straightforward manner (see, e.g., Ref. [15]) by diagonalizing the full Hamiltonian in the truncated Fock-state basis and solving the Schrödinger equation.[1] Figure 4.1a shows the population dynamics of the signal and idler modes, for different initial atom numbers N_0 and the quadratic Zeeman term tuned to the phase-matching condition $q = gN_0$. Setting $q = 0$ eliminates the Zeeman shift and we observe (grey solid line) significantly slowed dynamics due to phase mismatch. For reference, we also mark the experimental measurement time of Ref. [2], $\tau' = 0.0073$, corresponding to the reported value of the squeezing parameter $r \equiv N_0\tau' \simeq 2$ [25], evaluated for $N_0 = 275$.

---

[1]This method can be easily implemented for modelling the pump mode being initially either in a pure Fock state or in a coherent state. For the large values of N_0 and relatively small time durations considered in this paper, the two alternatives give very similar results; accordingly, we restrict ourselves to presenting the results only for the coherent initial state.

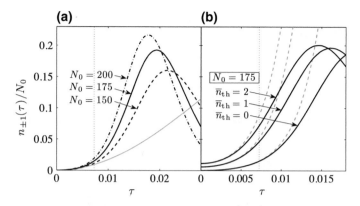

Fig. 4.1 a Fractional population $n_{\pm1}(\tau)/N_0$ of the signal/idler modes [where $n_{\pm1}(\tau) \equiv \langle \hat{a}_{\pm1}^{\dagger}(\tau)\hat{a}_{\pm1}(\tau)\rangle$] as a function of the dimensionless time τ, for vacuum initial state and different initial number of atoms in the pump mode, N_0. The quadratic Zeeman term is phase-matched to $q = gN_0$ in all cases, except for the *grey solid line* which is shown for comparison for $q = 0$ and $N_0 = 175$. The *vertical dotted line* indicates the measurement time $\tau' = 0.0073$ used in Ref. [2]. **b** Same as in (**a**) but with thermally seeded populations in the signal/idler modes (assumed to be equal to each other), for $N_0 = 175$. The *grey dashed lines* show the analytic predictions in the undepleted pump approximation

We next analyze the case of an initial thermal seed in the signal/idler modes, with an equal average number of atoms $\bar{n}_{th}$ in both modes. To simulate the dynamics in this case, we use the truncated Wigner method (Ref. [26] gives simple prescriptions on how to model various initial states in the Wigner representation). Figure 4.1b illustrates that the presence of the thermal seed accelerates population growth, however, it does not significantly effect the maximal depletion of the BEC. The numerical results in Fig. 4.1b are compared with the analytic predictions of the simple model of parametric down-conversion in the undepleted pump approximation, $n_{\pm1}(\tau) = \sinh^2(N_0\tau)[1 + 2\bar{n}_{th}] + \bar{n}_{th}$ (see Appendix E for full analytic solutions). As expected, we find good agreement between the numerical and analytic results in the short-time limit. We also conclude that as far as the mode populations are concerned, the experimental measurement time $\tau' = 0.0073$ is not too far from the regime of validity of the simple analytic model, at least for $\bar{n}_{th} \lesssim 2$. This conclusion, however, cannot necessarily be carried through to other observables, such as entanglement measures analysed below.

4.3.2 EPR Entanglement

Central to this paper is an investigation into the possible demonstration of the EPR paradox as outlined in Ref. [2]. In the context of continuous-variable entanglement, this is equivalent to the seeming violation of the Heisenberg uncertainty relation for inferred quadrature variances [12, 15]. In the normalised form this EPR entanglement criterion can be written as

$$\Upsilon_j = \frac{\Delta_{\text{inf}}^2 \hat{X}_j \Delta_{\text{inf}}^2 \hat{Y}_j}{(1 - \langle \hat{a}_j^\dagger \hat{a}_j \rangle / \langle \hat{b}_j^\dagger \hat{b}_j \rangle)^2} < 1, \tag{4.6}$$

where the optimal[2] inferred quadrature variance for $\hat{X}_j$ (and similarly for $\hat{Y}_j$) is given by [12]

$$\Delta_{\text{inf}}^2 \hat{X}_j = \langle (\Delta \hat{X}_j)^2 \rangle - \frac{\langle \Delta \hat{X}_i \Delta \hat{X}_j \rangle^2}{\langle (\Delta \hat{X}_i)^2 \rangle}, \tag{4.7}$$

with $\Delta \hat{X}_j \equiv \hat{X}_j - \langle \hat{X}_j \rangle$ and $i, j = \pm 1$. The generalized quadrature operators are defined as $\hat{X}_j(\theta) = (\hat{a}_j^\dagger \hat{b}_j e^{i\theta} + \hat{b}_j^\dagger \hat{a}_j e^{-i\theta}) / \langle \hat{b}_j^\dagger \hat{b}_j \rangle^{1/2}$ [15], where the operator $\hat{b}_j$ represents the local oscillator field required for homodyne detection of the quadratures and we denote $\hat{X}_j = \hat{X}_j(\pi/4)$ and $\hat{Y}_j = \hat{X}_j(3\pi/2)$. Choosing this pair of canonically conjugate quadratures maximises the correlation (anti-correlation) between them, defined as $C = \langle \hat{X}_i(\theta) \hat{X}_j(\theta) \rangle / [\langle \hat{X}_i(\theta)^2 \rangle \langle \hat{X}_j(\theta)^2 \rangle]^{1/2}$, thus minimizing the inferred quadrature variance.

Our choice of generalized quadrature operators [15] varies from the standard form, $\hat{X}_j(\theta) = \hat{a}_j e^{-i\theta} + \hat{a}_j^\dagger e^{i\theta}$ [25], as it does not assume a perfectly coherent, strong local oscillator. Instead, it takes into account the fact that the local oscillator is derived, just before the measurement time, from the partially depleted and already incoherent pump mode [2]. When measuring these quadratures the pump mode is split into two local oscillators by an atomic beam-splitter [15] (for instance a rf $\pi/2$ pulse), in which the output is given by $\hat{b}_{\pm 1} = (\hat{a}_0 \pm \hat{a}_{\text{vac}})/\sqrt{2}$, where $\hat{a}_{\text{vac}}$ represents the vacuum entering the empty port of the beam-splitter. This is slightly different to the method used in Ref. [2], where an atomic three-port beam-splitter is used to measure relevant quadratures.

Phase accrued due to $\hat{H}_{\text{el}} + \hat{H}_Z$ leads to a drifting in the phase relation between the local oscillator and the signal/idler modes. This means that our original quadrature choice of $\hat{X}_j(\pi/4)$ and $\hat{X}_j(3\pi/2)$ may not measure the optimal violation of the EPR criterion. By minimizing this criterion as a function of phase the optimal choice of quadratures becomes $\hat{X}_j(\theta_0(\tau))$ and $\hat{X}_j(\theta_0(\tau) + \pi/2)$, where $\theta_0(\tau)$ is the optimal local oscillator phase relative to the signal/idler modes.

In Fig. 4.2a we show the results of calculation of the phase-optimized EPR entanglement parameter Υ (with $\Upsilon_{-1} = \Upsilon_1 \equiv \Upsilon$ due to the symmetry of the signal/idler modes) for the signal/idler modes initially in a vacuum state. We see that strong EPR entanglement ($\Upsilon < 1$) can be achieved for a large experimental time frame,

---

[2]The form of the inferred quadrature variance in Eq. (4.7) varies slightly from that used in Ref. [2], where the inferred quadrature variances are equivalent to measurements of $\Delta_{\text{inf}}^2 \hat{X}_2 = \Delta^2(\hat{X}_1 - \hat{X}_2)$ and $\Delta_{\text{inf}}^2 \hat{Y}_2 = \Delta^2(\hat{Y}_1 + \hat{Y}_2)$ [15]. This choice is different in that it does not give the optimal violation of Eq. (4.6) [12], however, in the parameter regime we consider the difference between the choices of inferred quadratures is not qualitatively significant.

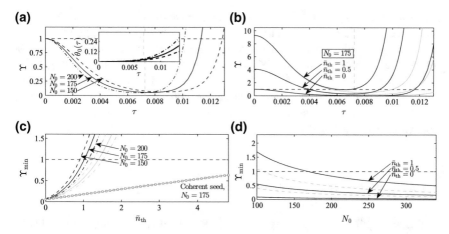

Fig. 4.2 **a** Evolution of the EPR entanglement parameter Υ for the same situation as in Fig. 4.1a. The EPR criterion corresponds to $\Upsilon < 1$ (*dashed horizontal line*). The *inset* shows the evolution of the optimal phase angle of the local oscillator $\theta_0(\tau)$ for each N_0. **b** Evolution of Υ for thermally seeded signal/idler modes and $N_0 = 175$. The experimental measurement time $\tau' = 0.0073$ is shown in (**a**) and (**b**) as a *vertical dotted line*. The respective *grey lines* are the analytic predictions from the undepleted pump model. **c** Time-optimized EPR parameter $\Upsilon_{\min}$ as a function of $\bar{n}_{\text{th}}$, for different N_0. The respective *grey lines* are the analytic predictions from the undepleted pump model. The *grey line* with squares shows $\Upsilon_{\min}$ for $N_0 = 175$, but assuming that the seeds are in a coherent state (sharing initially the same phase as the pump mode) with average populations of $|\alpha_{\pm 1}(0)|^2 = \bar{n}_{\text{th}}$. **d** Same as in (**c**), but as a function of N_0, for three different thermal seeds $\bar{n}_{\text{th}}$

up to $\tau \simeq 0.01$; more specifically, we predict suppression of the optimal EPR entanglement of at least 90 % below unity for all relevant total atom numbers (ranging from 150 to 200) at $\tau' = 0.0073$. Unlike the simple undepleted pump model, which predicts $\Upsilon = \cosh^{-2}(2N_0\tau)$ and hence indefinite suppression of the EPR criterion [25], EPR entanglement in the full model is eventually lost due to a combination of back-conversion ($|+1\rangle + |-1\rangle \rightarrow |0\rangle + |0\rangle$) and the loss of coherence in the pump mode.

Our results predict that a strong EPR violation should have been observed if the signal and idler modes were indeed generated from an initial vacuum state. In light of this and the large error margin of the experimental result in Ref. [2], which thus cannot conclusively demonstrate the existence or non-existence of EPR entanglement, we now discuss the possible presence of stray or thermally excited atoms in the signal/idler modes and the effects such seeding can have on entanglement and particularly the EPR criterion. The results of calculation of the EPR entanglement parameter Υ for an initial thermal seed of $\bar{n}_{\text{th}}$ in both modes are shown in Fig. 4.2b–d. We find the introduction of a thermal seed reduces the strong correlation between the signal and idler modes, leading to an eventual loss of EPR entanglement. For an initial number of atoms in the pump mode ranging between 150 to 200, EPR entanglement is lost already for $\bar{n}_{th} \simeq 1$. Direct experimental detection of stray atoms at such a low population level is beyond the current resolution of absorption imaging techniques [2]. More generally, our numerical results show that the maximum $\bar{n}_{th}$ that can be tolerated while preserving the EPR entanglement scales as $(\bar{n}_{th})_{max} \sim 0.06 N_0^{11/20}$ in the range of $100 \lesssim N_0 \lesssim 400$ (see Appendix E for further discussion). For comparison, seeding the signal and idler modes with a coherent state [17] of similar population does not have such a dramatic effect on EPR entanglement [see the grey line with squares in Fig. 4.2c].

4.3.3 Quadrature Squeezing and Inseparability

To further highlight the high sensitivity of EPR entanglement to initial thermal noise we contrast it with two other weaker measures of the nonclassicality of the state: two-mode quadrature squeezing and intermode entanglement in the sense of inseparability, which were the main focus of Ref. [2]. The two-mode quadrature variances are defined as $\hat{X}_{\pm}(\theta) = \hat{X}_1(\theta) \pm \hat{X}_{-1}(\theta)$, with $\Delta^2 \hat{X}_-(\theta) < 2$ corresponding to two-mode squeezing [25], *i.e.*, suppression of fluctuations below the level dictated by a minimum uncertainty state. We plot the results of our numerical calculations of quadrature variances in Fig. 4.3a. From these results we observe that the measurements of Ref. [2] do not agree with the amplitude of the oscillation that we find for an initial vacuum state (solid lines) or a small coherent seed (dot-dashed line). Rather they suggest the presence of a small thermal seed of $\bar{n}_{th} \simeq 1$ (dashed lines), although for definitive differentiation of initial thermal or coherent populations further experimental measurements with reduced error margins are required. Further, calculation of the minimum of $\Delta^2 \hat{X}_-$ [Fig. 4.3b, c] highlights that two-mode squeezing is preserved for thermal seed populations up to $\bar{n}_{th} \simeq 1.7$, which is consistent with our interpretation of the measurements reported in Ref. [2].

Next we define the sum of single-mode quadrature variances as $\sum \Delta_1^2 = 2(\Delta^2 \hat{X}_1 + \Delta^2 \hat{Y}_1)$ and the sum of two-mode quadrature variances, $\sum \Delta_2^2 = \Delta^2 \hat{X}_- + \Delta^2 \hat{Y}_+$. (Following the treatment of Ref. [2] we calculate the single-mode quadrature variances with the standard definition of quadratures, $\hat{X}_j(\theta) = \hat{a}_j e^{-i\theta} + \hat{a}_j^\dagger e^{i\theta}$.) Inseparability of the produced $m_F = \pm 1$ pair-entangled state is equivalent to $\sum \Delta_2^2 / \sum \Delta_1^2 < 1$ [27]. Figure 4.4a, b demonstrate that this measure of entanglement is far less sensitive to the presence of a thermal seed in comparison to the stronger criterion of EPR entanglement. Also, unlike the EPR criterion, this inseparability measure does not significantly differentiate between coherent and thermal seeding.

Fig. 4.3 **a** Two-mode quadrature variances $\Delta^2 \hat{X}_\pm (\theta)$ at $\tau' = 0.0073$ as functions of the local oscillator phase angle $\theta - \theta_0$, for vacuum (*solid lines*) and thermally seeded (*dashed lines*) signal/idler modes; $N_0 = 175$ in both cases. We also include a calculation of $\Delta^2 \hat{X}_- (\theta)$ for comparable coherent seed (*dot-dashed line*), $|\alpha_{\pm 1}(0)|^2 = 1$, which is almost indistinguishable from the vacuum case. **b** Time-optimized minimum of $\Delta^2 \hat{X}_- (\theta_0)$ as a function of $\bar{n}_{th}$, for different N_0. The *grey line* with squares shows $\Delta^2 \hat{X}_- (\theta_0)$ for $N_0 = 175$, but assuming the seeds are in a coherent state with average populations of $|\alpha_{\pm 1}(0)|^2 = \bar{n}_{th}$. **c** Same as in (**b**), but as a function of N_0, for different $\bar{n}_{th}$

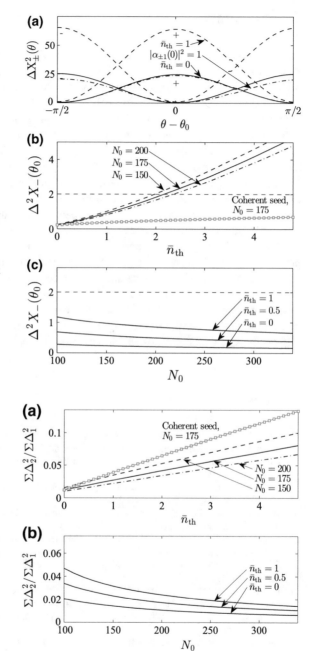

Fig. 4.4 **a** Time-optimized inseparability criterion for the quadrature entangled state, quantified via $\sum \Delta_2^2 / \sum \Delta_1^2 < 1$, as a function of $\bar{n}_{th}$, for different N_0. The *grey line* with squares shows $\sum \Delta_2^2 / \sum \Delta_1^2$ for $N_0 = 175$, but assuming the seeds are in a coherent state with average populations $|\alpha_{\pm 1}(0)|^2 = \bar{n}_{th}$. **b** Same as in (**a**), but as a function of N_0, for different $\bar{n}_{th}$

4.4 Summary

In conclusion, we have demonstrated that for an initial vacuum state in the signal/idler modes a strong suppression of the EPR criterion can be achieved in the parameter regime of Ref. [2], most importantly including the experimental measurement time of $\tau' = 0.0073$. However, we also establish that the strength of EPR entanglement depends crucially on the nature of the initial spin-fluctuations. Specifically, we predict that for a pump mode of initially 150 to 200 atoms, a thermal initial seed of $\bar{n}_{th} \simeq 1$ is sufficient to rule out EPR entanglement. Weaker measures of entanglement, such as inseparability, are still possible to observe as these are far more robust to thermal noise. This implies that spin-changing collisions may still be a good source of entanglement even in the presence of large thermal effects, even though we may not be able to carry through the EPR arguments that confront the completeness of quantum mechanics and advocate for local hidden variable theories. Importantly, our results suggest that the measurement of this EPR criterion can serve as a sensitive probe of the initial state which triggers the pair production process, beyond measures employed in Ref. [17]. This understanding of the sensitivity of EPR entanglement to initial thermal noise will hopefully lead to refining of spin-mixing experiments towards demonstration of the EPR paradox with massive particles. We expect our findings to be also relevant to related proposals based on molecular dissociation [28, 29], condensate collisions [30–33], and optomechanical systems [34, 35].

References

1. Ou, Z.Y., Pereira, S.F., Kimble, H.J., Peng, K.C.: Realization of the Einstein-Podolsky-Rosen paradox for continuous variables. Phys. Rev. Lett. **68**, 3663–3666 (1992)
2. Gross, C., et al.: Atomic homodyne detection of continuous-variable entangled twin-atom states. Nature **480**, 219 (2011)
3. Schrodinger, E.: Discussion of probability relations between separated systems. Math. Proc. Camb. Philos. Soc. **31**, 555–563 (1935)
4. Giovannetti, V., Lloyd, S., Maccone, L.: Quantum-enhanced measurements: beating the standard quantum limit. Science **306**, 1330–1336 (2004)
5. Gross, C., Zibold, T., Nicklas, E., Estève, J., Oberthaler, M.K.: Nonlinear atom interferometer surpasses classical precision limit. Nature **464**, 1165 (2010)
6. Riedel, M.F., et al.: Atom-chip-based generation of entanglement for quantum metrology. Nature **464**, 1170 (2010)
7. Einstein, A., Podolsky, B., Rosen, N.: Can quantum-mechanical description of physical reality be considered complete? Phys. Rev. **47**, 777–780 (1935)
8. Bohr, N.: Can quantum-mechanical description of physical reality be considered complete? Phys. Rev. **48**, 696–702 (1935)
9. Bohm, D.: A suggested interpretation of the quantum theory in terms of "hidden" variables. I. Phys. Rev. **85**, 166–179 (1952)
10. Bohm, D., Aharonov, Y.: Discussion of experimental proof for the paradox of Einstein, Rosen, and Podolsky. Phys. Rev. **108**, 1070–1076 (1957)
11. Bell, J.S.: On the Einstein-Podolsky-Rosen paradox. *Phys. (N.Y.)* **1**, 195 (1964)
12. Reid, M.D.: Demonstration of the Einstein-Podolsky-Rosen paradox using nondegenerate parametric amplification. Phys. Rev. A **40**, 913–923 (1989)

13. Pu, H., Meystre, P.: Creating macroscopic atomic einstein-podolsky-rosen states from bose-einstein condensates. Phys. Rev. Lett. **85**, 3987–3990 (2000)
14. Duan, L.-M., Sørensen, A., Cirac, J.I., Zoller, P.: Squeezing and entanglement of atomic beams. Phys. Rev. Lett. **85**, 3991–3994 (2000)
15. Ferris, A.J., Olsen, M.K., Cavalcanti, E.G., Davis, M.J.: Detection of continuous variable entanglement without coherent local oscillators. Phys. Rev. A **78**, 060104 (2008)
16. Melé-Messeguer, M., Juliá-Díaz, B., Polls, A., Santos, L.: Thermal spin fluctuations in spinor Bose-Einstein condensates. Phys. Rev. A **87**, 033632 (2013)
17. Klempt, C., et al.: Parametric amplification of vacuum fluctuations in a spinor condensate. Phys. Rev. Lett. **104**, 195303 (2010)
18. Law, C.K., Pu, H., Bigelow, N.P.: Quantum spins mixing in spinor Bose-Einstein condensates. Phys. Rev. Lett. **81**, 5257–5261 (1998)
19. Pu, H., Law, C.K., Raghavan, S., Eberly, J.H., Bigelow, N.P.: Spin-mixing dynamics of a spinor Bose-Einstein condensate. Phys. Rev. A **60**, 1463–1470 (1999)
20. Chang, M.-S., Qin, Q., Zhang, W., You, L., Chapman, M.S.: Coherent spinor dynamics in a spin-1 Bose condensate. Nat. Phys. **1**, 111–116 (2005)
21. Kawaguchi, Y., Ueda, M.: Spinor Bose-Einstein condensates. Phys. Rep. **520**, 253–381 (2012)
22. Kronjäger, J., et al.: Evolution of a spinor condensate: Coherent dynamics, dephasing, and revivals. Phys. Rev. A **72**, 063619 (2005)
23. Bize, S., et al.: High-accuracy measurement of the ^{87}rb ground-state hyperfine splitting in an atomic fountain. Europhys. Lett. **45**, 219–223 (1999)
24. Kronjäger, J., Becker, C., Navez, P., Bongs, K., Sengstock, K.: Magnetically tuned spin dynamics resonance. Phys. Rev. Lett. **97**, 110404 (2006)
25. Walls, D.F., Milburn, G.J.: Quantum Optics, 2nd edn. Springer, Berlin (2008)
26. Olsen, M., Bradley, A.: Numerical representation of quantum states in the positive-P and Wigner representations. Opt. Commun. **282**, 3924–3929 (2009)
27. Raymer, M.G., Funk, A.C., Sanders, B.C., de Guise, H.: Separability criterion for separate quantum systems. Phys. Rev. A **67**, 052104 (2003)
28. Kheruntsyan, K.V., Olsen, M.K., Drummond, P.D.: Einstein-Podolsky-Rosen correlations via dissociation of a molecular Bose-Einstein condensate. Phys. Rev. Lett. **95**, 150405 (2005)
29. Kheruntsyan, K.V.: Matter-wave amplification and phase conjugation via stimulated dissociation of a molecular Bose-Einstein condensate. Phys. Rev. A **71**, 053609 (2005)
30. Ferris, A.J., Olsen, M.K., Davis, M.J.: Atomic entanglement generation and detection via degenerate four-wave mixing of a Bose-Einstein condensate in an optical lattice. Phys. Rev. A **79**, 043634 (2009)
31. Jaskula, J.-C., et al.: Sub-Poissonian number differences in four-wave mixing of matter waves. Phys. Rev. Lett. **105**, 190402 (2010)
32. Kheruntsyan, K.V., et al.: Violation of the Cauchy-Schwarz inequality with matter waves. Phys. Rev. Lett. **108**, 260401 (2012)
33. Kofler, J., et al.: Einstein-Podolsky-Rosen correlations from colliding Bose-Einstein condensates. Phys. Rev. A **86**, 032115 (2012)
34. Vitali, D., et al.: Optomechanical entanglement between a movable mirror and a cavity field. Phys. Rev. Lett. **98**, 030405 (2007)
35. Müller-Ebhardt, H., Miao, H., Danilishin, S., Chen, Y.: Quantum-state steering in optomechanical devices. arXiv:1211.4315 (2012)

Chapter 5
An Atomic SU(1, 1) Interferometer via Spin-Changing Collisions

In the previous chapters we have demonstrated how the two-mode squeezed vacuum state can be realized in a variety of systems and used to test foundational concepts of quantum mechanics such as Bell inequalities and the EPR paradox. At the centre of these tests is the presence of strong correlations between the two modes, particularly, phase-sensitive correlations. Another practical application of these correlations is in the field of quantum metrology. In particular, when the two-mode squeezed vacuum state is used as the input into a suitable two-mode interferometer it enables interferometric sensitivity below the classical shot noise limit.

Here, we specifically investigate this application in the context of a SU(1, 1) interferometer. Distinct to other interferometers which use passive elements (such as optical beam-splitters) to probe the phase-sensitive correlations, we use the two-mode squeezing process itself as a form of active 'nonlinear' beam-splitter. The SU(1, 1) scheme can lead to benefits relative to passive interferometers, such as increased robustness to imperfect detection [1]. Our theoretical analysis is specifically focused on the realization of an atomic SU(1, 1) interferometer via spin-changing collisions in a spinor BEC, motivated by the recent experimental work of Refs. [2] and [3].

5.1 Framework of a Two-Mode Interferometer

To begin, we outline the theoretical formalism of parameter estimation in the context of two-mode interferometry, from which the SU(1, 1) interferometer naturally emerges. The simplest description of a two-mode interferometer is illustrated in Fig. 5.1. An input state, $|\psi_{\text{in}}\rangle$, is acted upon by a unitary process $\hat{U}(\phi_1, \phi_2)$ which imparts a phase shift dependent upon the path, defined as [4]

$$\hat{U}(\phi_1, \phi_2) = \exp\left(i\hat{G}_1\phi_1 + i\hat{G}_2\phi_2\right), \tag{5.1}$$

© Springer International Publishing Switzerland 2016
R.J. Lewis-Swan, *Ultracold Atoms for Foundational Tests of Quantum Mechanics*, Springer Theses, DOI 10.1007/978-3-319-41048-7_5

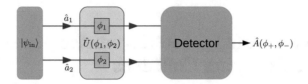

Fig. 5.1 Simplest two-mode interferometric scheme for parameter estimation of ϕ_1 and ϕ_2. An arbitrary two-mode input state $|\psi_{in}\rangle$ undergoes unitary evolution which imparts a phase shift on each mode. The final state is then measured by some arbitrary process with corresponding measurement signal $\hat{A}$, which is used in practice to estimate the values of the sum and difference, ϕ_+ and ϕ_-, of the phase shifts

where $\hat{G}_i$ for $i = 1, 2$ is known as the generator of the phase shift in each mode. In this chapter we will restrict our analysis to the simple case where $\hat{G}_i = \hat{n}_i$ and $\hat{n}_i = \hat{a}_i^\dagger \hat{a}_i$ is the occupation of each mode. Equation (5.2) can also be rewritten in the form [4]

$$\hat{U}(\phi_1, \phi_2) \equiv \hat{U}(\phi_+, \phi_-) = \exp\left(i\hat{G}_+\phi_+ + i\hat{G}_-\phi_-\right), \tag{5.2}$$

where we introduce the sum $\phi_+ = \phi_1 + \phi_2$ and difference $\phi_- = \phi_1 - \phi_2$ phases with generators $\hat{G}_\pm = \hat{n}_\pm/2$ respectively for $\hat{n}_\pm = \hat{n}_1 \pm \hat{n}_2$. After the phase shifts, the output state, $|\psi_{out}\rangle = \hat{U}(\phi_+, \phi_-)|\psi_{in}\rangle$, is measured by some detection process and a measurement signal $\hat{A}(\phi_+, \phi_-)$ is used to estimate the sum or differential phases.

From a purely theoretical point of view, the accuracy with which one can estimate the phase shift is limited by the quantum Cramer-Rao bound [5, 6],

$$(\Delta\phi_\pm)^2 \geq \frac{1}{\mathcal{F}_\pm} \tag{5.3}$$

where

$$\mathcal{F}_\pm = 4\left[\langle\partial_{\phi_\pm}\psi_{out}|\partial_{\phi_\pm}\psi_{out}\rangle - \langle\partial_{\phi_\pm}\psi_{out}|\psi_{out}\rangle\langle\psi_{out}|\partial_{\phi_\pm}\psi_{out}\rangle\right] \tag{5.4}$$

is the quantum Fisher information with respect to $\phi_\pm$ [4, 7]. However, in practice one can use the measurement signal $\hat{A}(\phi_+, \phi_-)$ and Gaussian error propagation to approximate the sensitivity as [6]

$$(\Delta\phi_\pm)^2 = \langle\Delta^2\hat{A}(\phi_+, \phi_-)\rangle \left/ \left|\frac{d\langle\hat{A}(\phi_+, \phi_-)\rangle}{d\phi_\pm}\right|^2\right. . \tag{5.5}$$

For an optimal choice of *estimator* $\hat{A}(\phi_+, \phi_-)$, Eq. (5.5) will saturate the quantum Cramer-Rao bound [6, 8].

Classically, the best sensitivity one can achieve is the shot-noise limit, which scales as $1/N$ where N is the number of quanta used in the interferometer (often referred to as the resource). This sensitivity is also known as the standard quantum limit

(SQL). However, by leveraging properties of quantum states such as entanglement and squeezing, one can show that the best sensitivity for a quantum system (utilizing a linear phase shift) scales as $1/N^2$ for $N \gg 1$, known as the Heisenberg limit [8]. Achieving this limit in practice requires an appropriate choice of state, such that $1/\mathcal{F}_\pm$ equals the Heisenberg limit, and the proper choice of measurement that saturates the quantum Cramer-Rao bound.

5.1.1 SU(1, 1) Interferometer

Having outlined the basic framework of two-mode interferometry, one may ask whether the strong correlations present in the two-mode squeezed vacuum state make it suitable for sub-shot noise interferometry. Taking it as the input state to the interferometer in Fig. 5.1, we can easily calculate the quantum Fisher information of the state for the sum and difference phase shifts:

$$\mathcal{F}_+ = \langle \Delta^2(\hat{n}_1 + \hat{n}_2) \rangle = n_s(n_s + 2), \tag{5.6}$$
$$\mathcal{F}_- = \langle \Delta^2(\hat{n}_1 - \hat{n}_2) \rangle = 0. \tag{5.7}$$

where $n_s \equiv \langle \hat{n}_+ \rangle$ is the mean sum population of the two-mode squeezed vacuum state entering the interferometer. One then sees that the two-mode squeezed vacuum state provides no useful information for an estimate of ϕ_- as $1/\mathcal{F}_-$ is undefined. The optimal uncertainty in an estimate of ϕ_+, however, is given by

$$(\Delta\phi_+)^2 = \frac{1}{n_s(n_s + 2)}, \tag{5.8}$$

which asymptotically scales as the Heisenberg limit for $n_s \gg 1$. These results are intuitive when one considers the number and phase fluctuations of the two-mode squeezed vacuum. As this state has strongly squeezed relative number difference fluctuations, $\langle \Delta^2(\hat{n}_1 - \hat{n}_2) \rangle = 0$, the relative phase of the two modes is undefined. Consequently, the state is insensitive to the differential phase ϕ_-. In contrast, the sum population exhibits strong fluctuations, $\langle \Delta^2(\hat{n}_1 + \hat{n}_2) \rangle = n_s(n_s + 2)$, which for the two-mode squeezed vacuum state implies that the sum phase of the two modes should be well defined (with respect to the phase of the pump mode), explaining the states strong sensitivity to changes in the sum phase ϕ_+.

Whilst this result indicates the two-mode squeezed vacuum state is a useful candidate for interferometry, the optimal measurement which will saturate the quantum Cramer-Rao bound is, however, not necessarily obvious in this case. For instance, a direct measurement of the sum population $\hat{n}_1 + \hat{n}_2$ or the difference $\hat{n}_1 - \hat{n}_2$ does not serve as a suitable interferometric signal. Following the work of Yurke et al. [9], it can be shown that for this state the optimal measurement scheme involves first applying another two-mode squeezing Hamiltonian and then making a measurement of the sum population $\hat{n}_+ = \hat{n}_1 + \hat{n}_2$. The sensitivity of the interferometer can then be calculated by applying the error propagation formula [Eq. (5.5)]:

$$(\Delta\phi_+)^2 = \langle\Delta^2\hat{n}_+\rangle \Big/ \left|\frac{d\langle\hat{n}_+\rangle}{d\phi_+}\right|^2 . \tag{5.9}$$

An interferometer combining this input state and measurement scheme is known as a SU(1, 1) interferometer and was first proposed by Yurke et al. [9]. Such an interferometer has only recently been realized for the first time in quantum optics [10], and an example is illustrated in Fig. 5.2a. It is composed of an initial two-mode squeezing process, which for illustrative purposes we take to be spontaneous parametric down-conversion of photons in a $\chi^{(2)}$ nonlinear medium, to produce the two-mode squeezed vacuum state. This is followed by separate linear phase shifts, ϕ_1 and ϕ_2, on each mode before the photons are passed back through another $\chi^{(2)}$

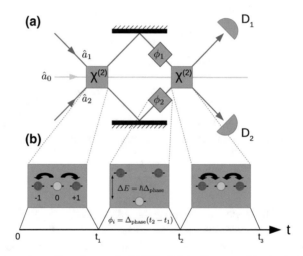

Fig. 5.2 **a** Schematic outline of an optical SU(1, 1) interferometer. The 'active' beam-splitters are realized by a $\chi^{(2)}$ nonlinear medium, which is pumped by photons from a strong coherent field to realize the archetypal two-mode squeezing Hamiltonian (see Eq. (1.8) of Sect. 1.4.1 for details). After the nonlinear medium the effective input state is a two-mode squeezed vacuum, which undergoes a linear phase shift dependent on the path through the interferometer. To estimate the sum phase $\phi_+ = \phi_1 + \phi_2$ a second nonlinear medium is placed before the detectors (D_1 and D_2) which measure the respective mode populations. **b** Equivalent realization in a spinor BEC. The active beam-splitters are realized by spin-changing collisions (which are run from $t = 0$ until $t = t_1$), and which in the first sequence transfer pairs of atoms from the $m_F = 0$ (*pump*) mode to the $m_F = \pm 1$ modes (*sidemodes*). The linear phase shift can be realized by increasing the Zeeman shift such that the splitting of the energy levels is $\Delta_{\text{phase}} = q_{\text{phase}} + gN_0$. This halts the spin-changing collisions (at time $t = t_1$) and leads to a phase accrual of $\phi_i = \Delta_{\text{phase}}(t_2 - t_1)$ (where the phase is accrued from $t = t_1$ until $t = t_2$) in each sidemode with respect to the pump phase. Depending on the phase relation between pump and sidemodes, the second period of spin-changing collisions (which occurs from $t = t_2$ until $t = t_3$) may transfer atoms from the pump to the sidemodes or vice versa

nonlinear medium. Depending on the sum phase relation between the signal and idler modes and the pump beam, the second nonlinear medium may continue down-conversion of photons into the signal and idler modes or the reverse will occur, wherein the photons from the signal and idler beams recombine into the pump mode by the process of second harmonic generation. The sum population of the signal and idler modes, which will be strongly dependent on the sum phase, is then measured at the output ports of the interferometer. Comparing to the common Mach-Zender interferometer, we see that the passive beam-splitter elements are replaced by the $\chi^{(2)}$ nonlinear mediums, which can be thought of as active 'nonlinear' beam-splitters.

5.2 Theoretical Analysis of an Atomic SU(1, 1) Interferometer

A practical candidate for realization of an atomic SU(1, 1) interferometer is a spinor BEC [1, 11, 12], in which the two-mode squeezing Hamiltonian is realized by coherent spin-changing collisions. Spinor BECs are a desirable candidate for this interferometer as they present a clean, isolated system with comparatively small loss inside the interferometer [1] and relatively simple control of the spin-changing collisions. Furthermore, by preparing the system appropriately one can freeze the spatial dynamics of condensate and isolate the relevant evolution of the system to the spin degree of freedom.

Our theoretical analysis is motivated by an experimental setup identical to that discussed in Chap. 4, and thus the same approximations may be made. For clarity, we deal with a mesoscopic ^{87}Rb condensate (containing 250–550 atoms) in a tight trapping potential, such that we may invoke the single mode approximation. The effective Hamiltonian governing the spin-changing collisions (which is equivalent for $F = 1$ and $F = 2$ systems in the short time limit) is identical to that of Chap. 4, however we rewrite it here for convenience. It can be broken into the form

$$\hat{H} = \hat{H}_{\text{inel}} + \hat{H}_{\text{el}} + \hat{H}_{\text{Z}}, \tag{5.10}$$

where

$$\hat{H}_{\text{inel}} = \hbar g(\hat{a}_0^\dagger \hat{a}_0^\dagger \hat{a}_{-1} \hat{a}_1 + \hat{a}_1^\dagger \hat{a}_{-1}^\dagger \hat{a}_0 \hat{a}_0) \tag{5.11}$$

$$\hat{H}_{\text{el}} = \hbar g(\hat{n}_0 \hat{n}_1 + \hat{n}_0 \hat{n}_{-1}), \tag{5.12}$$

$$\hat{H}_{\text{Z}} = \hbar q \left(\hat{n}_1 + \hat{n}_{-1} \right), \tag{5.13}$$

are the inelastic (spin-changing) collision, elastic collision and Zeeman contributions respectively. The coupling constant g is associated with s-wave collisions (see Chap. 4 for more details) and we have neglected the linear Zeeman shift as $\hat{n}_1 - \hat{n}_{-1}$ is a conserved quantity.

The working principle of the atomic interferometer can be broken into three stages [11], analogous to the optical interferometer, and is illustrated in Fig. 5.2b. Firstly, a BEC is prepared purely in the $(F, m_F) = (2, 0)$ state at $t = 0$ and evolves according to $\hat{H}$ until time t_1 ($\equiv t_{evo}$), during which time spin-changing collisions occur and the two-mode squeezed vacuum state is generated in the $m_F = \pm 1$ sidemodes. Next, between t_1 and t_2 the inelastic spin-changing collisions are halted by either: (i) coherently transferring the BEC from the $(F, m_F) = (2, 0)$ state such that atom pairs are no longer produced in the $(F, m_F) = (2, \pm 1)$ sidemodes, by for example shifting the BEC to the $F = 1$ hyperfine state, or (ii) increasing the quadratic Zeeman shift between the pump and sidemodes (to the value q_{phase}) sufficiently such that the spin-changing process becomes far off-resonant. We assume that in either case the inelastic spin-changing collisions characterized by $\hat{H}_{inel}$ can be neglected, such that the system evolves according to the Hamiltonian $\hat{H}' = \hat{H}_{el} + \hat{H}_Z$ until time $t_2 = t_1 + t_{phase}$ and accrues a linear phase shift $\phi_1 = \phi_2 = \phi(t_{phase})$ in the $m_F = \pm 1$ modes dependent upon the duration t_{phase}. It is assumed that the phase of the pump mode is fixed throughout this period as it is unaffected by the Zeeman shift. Finally, the spin-changing collisions are restarted, by either returning the BEC to the $(F, m_F) = (2, 0)$ state or returning the Zeeman shift to the initial value q, and the system evolves according to $\hat{H}$ until time $t_3 = t_2 + t_{evo}$, for a duration identical to the first sequence. A measurement of the sum population of the $m_F = \pm 1$ states then allows the construction of the interferometric signal with which to estimate the phase shift $\phi_+ \equiv 2\phi$.

5.2.1   Ideal Solution in the Undepleted Pump Approximation

Simple analytic results can be found for the interferometer by invoking the undepleted pump approximation, previously discussed in Sect. 1.4.3, wherein we assume that the $m_F = 0$ mode is initially a strong coherent state of amplitude $\alpha_0 = \sqrt{N_0}$ (chosen to be real without loss of generality) which does not change in time. We then make the replacement $\hat{a}_0 \rightarrow \alpha_0$ and the Hamiltonian simplifies to

$$\hat{H} = \hbar g N_0 \left(\hat{a}_1^\dagger \hat{a}_{-1}^\dagger + \hat{a}_1 \hat{a}_{-1} \right) + \hbar \left(g N_0 + q \right) \left(\hat{n}_1 + \hat{n}_{-1} \right). \tag{5.14}$$

Similarly, when the spin-changing collisions are halted, the system evolves according to:

$$\hat{H}' = \hbar \left(g N_0 + q \right) \left(\hat{n}_1 + \hat{n}_{-1} \right). \tag{5.15}$$

The dynamics of the system can most readily be solved in the Heisenberg picture. Evolution with respect to $\hat{H}$ is described by the Heisenberg equations of motion for the creation operators,

$$\frac{d\hat{a}_{\pm 1}}{dt} = -i g N_0 \hat{a}_{\mp 1}^\dagger - i(g N_0 + q)\hat{a}_{\pm 1}. \tag{5.16}$$

The pair of coupled operator equations can be exactly solved in this approximation and written in matrix form,

$$\begin{bmatrix} \hat{a}_1(t) \\ \hat{a}^\dagger_{-1}(t) \end{bmatrix} = U_{\text{evo}}(t) \begin{bmatrix} \hat{a}_1(0) \\ \hat{a}^\dagger_{-1}(0) \end{bmatrix}, \tag{5.17}$$

where $\hat{a}_1(0)$ and $\hat{a}^\dagger_{-1}(0)$ are given by the initial conditions of the system. The evolution matrix has the form

$$U_{\text{evo}}(t) = \begin{bmatrix} \alpha_{\text{evo}}(t) & \beta_{\text{evo}}(t) \\ [\beta_{\text{evo}}(t)]^* & [\alpha_{\text{evo}}(t)]^* \end{bmatrix}, \tag{5.18}$$

for

$$\alpha_{\text{evo}} = \cosh\left(\sqrt{(gN_0)^2 - \Delta^2}t\right) + i\frac{(gN_0) + q}{\sqrt{(gN_0)^2 - \Delta^2}}\sinh\left(\sqrt{(gN_0)^2 - \Delta^2}t\right) \tag{5.19}$$

$$\beta_{\text{evo}} = -i\frac{(gN_0)}{\sqrt{(gN_0)^2 - \Delta^2}}\sinh\left(\sqrt{(gN_0)^2 - \Delta^2}t\right), \tag{5.20}$$

and $\Delta = q + gN_0$. The evolution under $\hat{H}'$ is trivially solved in a similar manner and can also be cast in matrix form as

$$\begin{bmatrix} \hat{a}_1(t) \\ \hat{a}^\dagger_{-1}(t) \end{bmatrix} = U_{\text{phase}} \begin{bmatrix} \hat{a}_1(0) \\ \hat{a}^\dagger_{-1}(0) \end{bmatrix}, \tag{5.21}$$

where

$$U_{\text{phase}}(t) = \begin{bmatrix} e^{-i\phi(t)} & 0 \\ 0 & e^{i\phi(t)} \end{bmatrix}, \tag{5.22}$$

and $\phi(t) \equiv \Delta_{\text{phase}}t$ is the phase shift accrued in the $m_F = \pm 1$ modes due to the combined quadratic Zeeman and mean-field shifts during the holding period, characterized by $\Delta_{\text{phase}} = q_{\text{phase}} + gN_0$. We note that irrespective of how the spin-changing collisions are halted [(i.e. by increasing the Zeeman shift or coherently transferring the pump atoms from $(F, m_F) = (2, 0)$ to $(F, m_F) = (1, 0)$], such that the evolution is characterised by $\hat{H}'$, we choose the same quadratic Zeeman shift q_{phase} during the holding period from $t = t_2$ to $t = t_3$.

The solution of the full interferometer can then be found by solving the dynamics for each stage individually and using the sequential solutions for $\hat{a}_{\pm 1}(t)$ as the initial condition for the following period of evolution to give:

$$\begin{bmatrix} \hat{a}_1(t_3) \\ \hat{a}^\dagger_{-1}(t_3) \end{bmatrix} = U_{\text{evo}}(t_{\text{evo}})U_{\text{phase}}(t_{\text{phase}})U_{\text{evo}}(t_{\text{evo}}) \begin{bmatrix} \hat{a}_1(0) \\ \hat{a}^\dagger_{-1}(0) \end{bmatrix}. \tag{5.23}$$

where t_3 corresponds to the end of the interferometer [as illustrated in Fig. 5.2b].

At the output of the interferometer one finds the mean sum population as a function of $\phi_+ = 2\phi(t_{\text{phase}})$ is:

$$\langle \hat{n}_+(\phi_+; t_3) \rangle = 2 \left[\lambda n_s \sin(\phi_+/2) + \gamma \cos(\phi_+/2) \right]^2, \tag{5.24}$$

where $\lambda = \Delta/(gN_0)$, $\gamma = \sqrt{n_s \left[2 + (1 - \lambda^2)n_s \right]}$ and

$$n_s = \langle \hat{n}_1(t_1) + \hat{n}_{-1}(t_1) \rangle = 2\sinh^2(\sqrt{(gN_0)^2 - \Delta^2} t_1)/(1 - \lambda^2), \tag{5.25}$$

is the mean sum population of the $m_F = \pm 1$ sidemodes prepared in the two-mode squeezed vacuum state after the first period of spin-changing collisions. Importantly, n_s is taken to be the relevant quantity when defining the resource of the interferometer.

The variance of the sum population can be calculated in a similar fashion from Eq. (5.23), giving

$$\langle \Delta^2 \hat{n}_+(\phi_+; t_3) \rangle = \langle n_+(\phi_+; t_3) \rangle \left[2 + \langle n_+(\phi_+; t_3) \rangle \right]. \tag{5.26}$$

The result of Eqs. (5.27) and (5.26) can be most readily understood in the case where $q = -gN_0$ and the Hamiltonian $\hat{H}$ reduces entirely to the spin-changing collisions (squeezing) component. For $\phi_+ = \pi$ the phase-relation between the pump and sidemodes is such that effectively the sign of the spin-changing collisions term is reversed. This means that rather than produce pairs from the pump into the sidemodes as per $|2, 0\rangle \rightarrow |2, 1\rangle + |2, -1\rangle$, the process is reversed $|2, 1\rangle + |2, -1\rangle \rightarrow |2, 0\rangle$ and the second period of spin-changing collisions effectively undoes the first and $\langle \hat{n}_+(\pi; t_3) \rangle = 0$. Similarly, for $\phi_+ = 0$, the phase relation remains unchanged and the production of pairs in the $m_F = \pm 1$ sidemodes during the second period continues as if no interruption had occurred, growing exponentially such that $\langle \hat{n}_+(0; t_3) \rangle = n_s(n_s + 2)$.

The sensitivity of the interferometer can be constructed by substitution of Eqs. (5.24) and (5.26) into Eq. (5.9). One can show that optimal sensitivity occurs near:

$$\phi_+^{\text{opt}} = m\pi - \text{atan} \left(\frac{\gamma}{\lambda n_s} \right), \tag{5.27}$$

for $m \in \mathbb{Z}$. In practice, at exactly ϕ_+^{opt} we find $(\Delta\phi_+)^2$ is undefined, as from inspection of Eq. (5.24) we see that the working point of the interferometer is in fact a dark fringe [i.e minima of $\langle \hat{n}_+(\phi_+; t_3) \rangle$] wherein $\langle \hat{n}_+(\phi_+^{\text{opt}}; t_3) \rangle = 0$ and $\langle \Delta^2 \hat{n}_+(\phi_+^{\text{opt}}; t_3) \rangle = 0$. However, arbitrarily close to this point the sensitivity is found to limit to

$$\left(\Delta^2 \phi_+ \right)_{\text{opt}}^2 = \frac{1}{n_s(n_s + 2)}, \tag{5.28}$$

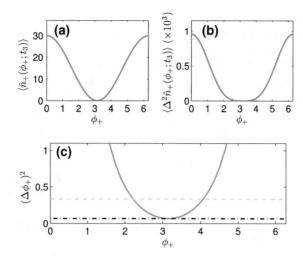

Fig. 5.3 **a** Plot of sum population $\langle \hat{n}_+(\phi_+; t_3) \rangle$ at output of interferometer for example case with $n_s = 3$ and $q = -gN_0$. **b** Variance of sum population $\langle \Delta^2 \hat{n}_+(\phi_+; t_3) \rangle$ for the same case. **c** Sensitivity $(\Delta\phi_+)^2$ (*blue solid line*) compared to standard quantum limit (*dashed grey line*) and Heisenberg limit (*dot-dashed black line*). The working point of the interferometer, i.e. optimal sensitivity, corresponds to a dark fringe in the mean sum population such that $\langle \hat{n}_+(\phi_+^{\text{opt}}; t_3) \rangle = 0$

which saturates the quantum Cramer-Rao bound. This implies that $\hat{n}_+$, combined with the second period of squeezing, is an optimal measurement in the limit of the undepleted pump approximation.

For illustration, we plot example results in Fig. 5.3 for Eqs. (5.24), (5.26) and (5.28). We choose parameters similar to the experiment of Refs. [3, 11] with $n_s = 3$ and $q = -gN_0$.

5.2.2 Effects of Off-Resonant Collisions

The working principle of the atomic SU(1, 1) interferometer crucially relies on the ability to 'switch-off' the spin-changing collisions during the phase shift component of the scheme [i.e. between t_1 and t_2 in Fig. 5.2b]. In the quantum optics realization, wherein one utilizes optical spontaneous parametric down-conversion as in Fig. 5.2a, the squeezing Hamiltonian is trivially stopped as the pump mode will leave the $\chi^{(2)}$ medium and enter free-space. As outlined in Sect. 5.2, the closest analog to this for the atomic case is to coherently transfer the BEC from the $(F, m_F) = (2, 0)$ to the $F = 1$ hyperfine state [usually directly to $(F, m_F) = (1, 0)$], which prohibits the spin-changing collision process transferring atoms between the $m_F = 0$ and $m_F = \pm 1$ modes in the $F = 2$ hyperfine state. A second method which can be considered is to increase the quadratic Zeeman energy shift between the pump and sidemodes, such that the spin-changing process between $m_F = 0$ and $m_F = \pm 1$

states is sufficiently far off-resonance that these terms in the Hamiltonian can be completely neglected. In such a case, we make the approximation that $\hat{H}_{inel}$ can be neglected and the system evolves according to $\hat{H}' = \hat{H}_{el} + \hat{H}_Z$. However, ensuring the Zeeman shift is sufficiently large that this is a good assumption is not necessarily a trivial task in experimental conditions. In the following, we characterize the validity of this approximation in the context of recent experimental work undertaken to realize an atomic SU(1, 1) interferometer [3]. Furthermore, we highlight two important consequences for the interferometric scheme when this assumption is not well founded.

Rather than explicitly assuming that the inelastic spin-changing collision terms can be neglected from t_1 until t_2, we may model the system with the full Hamiltonian $\hat{H} = \hat{H}_{inel} + \hat{H}_{el} + \hat{H}_Z$ (with the Zeeman shift given by q_{phase}) wherein we include these terms. In this case, we can solve the dynamics of the system as previously in the Heisenberg picture, however, in Eq. (5.23) we replace the idealized version of U_{phase} [Eq. (5.22)] with

$$U_{phase}(t) = \begin{bmatrix} \alpha_{phase}(t) & \beta_{phase}(t) \\ [\beta_{phase}(t)]^* & [\alpha_{phase}(t)]^* \end{bmatrix}, \tag{5.29}$$

where

$$\begin{aligned} \alpha_{phase}(t) = &\cosh\left(\sqrt{(gN_0)^2 - \Delta_{phase}^2}\,t\right) \\ &+ i\frac{(gN_0) + q}{\sqrt{(gN_0)^2 - \Delta_{phase}^2}}\sinh\left(\sqrt{(gN_0)^2 - \Delta_{phase}^2}\,t\right), \end{aligned} \tag{5.30}$$

$$\beta_{phase}(t) = -i\frac{(gN_0)}{\sqrt{(gN_0)^2 - \Delta_{phase}^2}}\sinh\left(\sqrt{(gN_0)^2 - \Delta_{phase}^2}\,t\right). \tag{5.31}$$

In the limit of $\Delta_{phase} \gg gN_0$, Eq. (5.29) collapses to the ideal form of Eq. (5.22).

With this form the mean sum occupation of the $m_F = \pm 1$ modes after the holding period (and before the second period of spin-changing collisions) is:

$$\langle \hat{n}_+(t_2) \rangle = n_s + \frac{2\sin(\phi)}{\sqrt{\lambda_{phase}^2 - 1}}\left[\epsilon\sin(\phi) + \gamma\cos(\phi)\right], \tag{5.32}$$

where $\epsilon = [1 + (1 - \lambda_{phase}\lambda)n_s]/\sqrt{\lambda_{phase}^2 - 1}$ and $\lambda_{phase} = \Delta_{phase}/(gN_0)$. The key observation from Eq. (5.32) is that the number of atoms in the side-modes during the accrual of the phase shift is no longer fixed, with an amplitude of oscillation given by $\sqrt{\gamma^2 + \epsilon^2}/\lambda_{phase}^2 - 1$. For $\Delta_{phase} \gg gN_0$ and $\lambda_{phase} \gg 1$ the amplitude vanishes $\sqrt{n_s(n_s + 2)}/|\lambda_{phase}| \to 0$ as expected.

In the experiment of Ref. [3], the Zeeman shift during the holding period gives a ratio $\lambda_{phase} \simeq 1.8$. We find that for the relevant atom number, $n_s = 3$, the population of the sidemodes inside the interferometer is predicted to oscillate with an amplitude of $\simeq 2.2$, which is a fluctuation of almost 75 % with respect to n_s.

This oscillatory atom number during the holding period presents a conceptual issue regarding how one defines the resource of the interferometer. Consequentially, it is difficult to define a SQL or Heisenberg limit for the interferometric scheme. In practice, as these oscillations cannot be completely removed, to approximate the atom number as constant a sufficient criteria would be to limit the oscillation amplitude to less than the minimum experimental error in measuring the population. For the experimental regime considered here this would require a two-fold increase in λ_{phase}.

A second and more fundamental issue arising from the inclusion of $\hat{H}_{inel}$ during the phase shift is that the canonical form of Eq. (5.2) is no longer valid. In particular, the scheme no longer corresponds to a linear phase shift where the generators are proportional only to $\hat{n}_1$ and $\hat{n}_{-1}$. This is not necessarily a terminal consequence for the scheme, as it has been shown that one can surpass the Heisenberg limit (for a linear phase shift) of an SU(2) interferometer with, for instance, an interferometer using a nonlinear phase shift where the generator is proportional to $\hat{n}^2$ [13]. However, this is an issue as we strive for an experimental realization of the SU(1, 1) interferometer as proposed by Yurke et al. [9], wherein the generator is given by $\hat{G}_+ = \hat{n}_+$ solely.

Experimentally, for the case of ^{87}Rb and other atoms with multiple hyperfine levels (i.e., $F \geq 2$), we can overcome these issues by taking advantage of the atom's hyperfine structure and coherently transfer the BEC from the $(F, m_F) = (2, 0)$ state to the $(F, m_F) = (1, 0)$ state to halt the spin-changing collisions. It should be noted that in the absence of the highly occupied pump mode, other processes in the $F = 2$ hyperfine state, such as those involving the $m_F = \pm 2$ modes, are also effectively halted. Furthermore, in the $F = 1$ state, the $(F = 1)$ coupling coefficient g' is sufficiently decreased such that $q_{phase} \gg g'N_0$ in this level also, implying that the spin-changing process $|1, 0\rangle \rightarrow |1, 1\rangle + |1, -1\rangle$ is strongly off-resonant and the pump mode does remains at fixed occupation from t_1 to t_2. By freezing the spin-changing dynamics with this transfer, the phase shift remains trivially linear and U_{phase} is given exactly by Eq. (5.22).

5.3 Conclusion

We have demonstrated that the entanglement and phase-sensitive correlations present in the atomic two-mode squeezed vacuum state have applications in quantum metrology. Specifically, we demonstrate how one can construct a SU(1, 1) interferometer where the two-mode squeezed vacuum is the optimal input state to obtain interferometric sensitivity at the Heisenberg limit. In particular, we have demonstrated that the pair-production process of spin-changing collisions in a spinor condensate is an excellent candidate to generate this state and realize an atomic SU(1, 1) interferometer. Treating the system in the undepleted pump approximation, we have derived generic results for the dynamics and phase sensitivity of the interferometer, which may also be relevant to other atom-optics systems [14]. We have also identified a key experimental difference to the quantum optics realization. Specifically, in the atom-optics scheme the $m_F = \pm 1$ states must be sufficiently detuned from the pump mode or isolated by transferal between hyperfine levels such that the spin-changing collisions are completely halted and the archetypal SU(1, 1) scheme is properly realized. In contrast, in an optical realization with spontaneous parametric down-conversion, this control is trivially realized by the finite length of the $\chi^{(2)}$ nonlinear medium. This theoretical understanding of the atom-optics system will hopefully lead to an experimental realization of an atomic SU(1, 1) interferometer for the first time [3].

References

1. Marino, A.M., Corzo Trejo, N.V., Lett, P.D.: Effect of losses on the performance of an SU(1,1) interferometer. Phys. Rev. A **86**, 023844 (2012)
2. Gross, C., et al.: Atomic homodyne detection of continuous-variable entangled twin-atom states. Nature **480**, 219 (2011)
3. Linnemann, D., Lewis-Swan, R.J., Strobel, H., Mussel, W., Kheruntstyan, K.V., Oberthaler, M.K.: Quantum-enhanced sensing based on time reversal of non-linear dynamics (2016). arXiv:1602.07505
4. Lang, M.D., Caves, C.M.: Optimal quantum-enhanced interferometry using a laser power source. Phys. Rev. Lett. **111**, 173601 (2013)
5. Braunstein, S.L., Caves, C.M.: Statistical distance and the geometry of quantum states. Phys. Rev. Lett. **72**, 3439–3443 (1994)
6. Wiseman, H.M., Milburn, G.J.: Quantum measurement and control. Cambridge University Press, Cambridge (2009)
7. Pezze, L., Smerzi, A.: Quantum theory of phase estimation. arXiv preprint arXiv:1411.5164 (2014)
8. Braunstein, S.L., Caves, C.M., Milburn, G.: Generalized uncertainty relations: theory, examples, and lorentz invariance. Ann. Phys. **247**, 135–173 (1996)
9. Yurke, B., McCall, S.L., Klauder, J.R.: SU(2) and SU(1,1) interferometers. Phys. Rev. A **33**, 4033–4054 (1986)
10. Hudelist, F., et al.: Quantum metrology with parametric amplifier-based photon correlation interferometers. Nat. Comm. **5** (2014)
11. Linnemann, D.: Realization of an SU(1,1) interferometer with spinor Bose-Einstein Condensates. Master's thesis, University of Heidelberg (2013)
12. Gabbrielli, M., Pezze, L., Smerzi, A.: Spin-mixing interferometry with Bose-Einstein condensates. arXiv preprint. arXiv:1503.08582 (2015)
13. Joo, J., et al.: Quantum metrology for nonlinear phase shifts with entangled coherent states. Phys. Rev. A **86**, 043828 (2012)
14. Bonneau, M., et al.: Tunable source of correlated atom beams. Phys. Rev. A **87**, 061603 (2013)

Chapter 6
On the Relation of the Particle Number Distribution of Stochastic Wigner Trajectories and Experimental Realizations

The Wigner function, or the Wigner quasi-probability distribution [1–5], has proven to be a versatile tool in understanding quantum mechanics. Firstly, by providing a complete representation of the quantum mechanical density operator in phase space, the Wigner function serves as the quantum moment-generating functional that allows the calculation of quantum mechanical expectation values of operators in the spirit of classical statistical physics. Secondly, the Wigner function has been extensively used in the so-called truncated Wigner approximation as a calculation technique for quantum dynamical simulations, most notably in the fields of quantum optics and ultracold atoms [6–20]. This latter utility follows from the possibility of converting the master equation for the quantum density operator into a generalised Fokker–Planck equation, which itself—for dissipationless systems and after truncation of third- and higher-order derivative terms (if any) [21]—acquires the form of a classical Liouville equation and can be cast as an equivalent set of (stochastic) c-number differential equations for the phase-space variables.

Despite the formal analogy of the evolution equation for the Wigner function to the Liouville equation for the classical probability distribution, the strict interpretation of the Wigner function as a true probability distribution fails as it attains negative values for certain quantum states. Furthermore, even when the Wigner function is strictly non-negative, its difference from a classical probability distribution stems from the fact that it is still constrained by the quantum mechanical uncertainty principle: it is a joint probability distribution for quantum mechanically *incompatible* observables and, therefore, cannot be regarded as having direct physical significance. In the truncated Wigner approximation, this constraint manifests itself through the fact that even though the c-number differential equations formally coincide with their classical deterministic counterparts, the quantum mechanical uncertainties are mimicked via random initial conditions that are sampled stochastically from the Wigner-function representation of the initial density matrix.

Given this understanding and constraining ourselves to problems involving a non-negative initial Wigner function—such that its non-negativity throughout the ensuing dynamics is either intrinsically preserved (such as for systems described by Hamiltonians that depend no-higher-than quadratically on creation or annihilation

© Springer International Publishing Switzerland 2016

R.J. Lewis-Swan, *Ultracold Atoms for Foundational Tests of Quantum Mechanics*, Springer Theses, DOI 10.1007/978-3-319-41048-7_6

field operators) or enforced by the truncated Wigner approximation [21, 22]—we address the question of whether and when the individual stochastic trajectories can be thought of as a faithful representation of the outcomes of individual experimental runs.[1] Even though this question has been discussed in the literature previously [8, 10, 20, 23–28], the answer appears to be far from trivial. For example, Blakie et al. make the remark that for highly occupied *'classical'* states, such as those near the critical transition to a Bose–Einstein condensate, *"it is plausible that single realizations of Wigner trajectories should approximately correspond to a possible outcome of a given experiment"*. Furthermore, the question seems to be heuristically posed; instead, we seek to address it in an operationally defined manner.

In this chapter, we investigate the connection between the outcomes of Wigner trajectories and experimental runs by comparing the respective particle number distributions; for simplicity we focus on treating single-mode problems. Experimentally the particle number distribution is measured by counting shots in which n quanta are detected, for instance photons hitting a detector, and corresponds to the true particle number distribution defined strictly via $P_n = |\langle n|\psi\rangle|^2$ where $n = 0, 1, 2, \ldots$, for a pure state $|\psi\rangle$. Similarly, for all positive Wigner functions $W_{|\psi\rangle}(\alpha)$, where α is the complex field amplitude, we can formally introduce an operationally well defined binned number distribution $\tilde{P}_n$ by calculating $n_i = |\alpha_i|^2 - 1/2$, where the index i indicates an individual trajectory (or equivalently individual samples appropriately taken from a known Wigner function), and sorting the continuous values into discrete bins such that $\tilde{P}_n$ is the probability to find $n - 1/2 \leq n_i < n + 1/2$. The subtraction of $1/2$ in the calculation of $n_i = |\alpha_i|^2 - 1/2$ can be thought of as representing the subtraction on average of half a quantum of noise (that has been added to the initial state to mimic quantum fluctuations), which is required in the calculation of the average mode occupation ($\langle \hat{n} \rangle = \langle \hat{a}^\dagger \hat{a} \rangle \equiv \langle \alpha^* \alpha \rangle_W - 1/2$, where $\hat{n}$ is the particle number operator, while $\hat{a}^\dagger$ and $\hat{a}$ are the creation and annihilation operators) using the Wigner function due to its correspondence to expectation values of symmetrically ordered operator products.

We find that the defining feature governing the interpretation of $\tilde{P}_n$ as a valid approximation to the true P_n is the smoothness and the broadness of the Wigner function relative to the oscillatory structure in $W_{|n\rangle}(\alpha)$. For thermal states, this criterion is in fact equivalent to high mean occupation of the mode, and therefore our findings confirm the heuristic assertion of Blakie et al. [23] that such an interpretation is valid for highly occupied 'classical' states. However, we also show—using an explicit counterexample which is for a highly squeezed coherent state (the Wigner function of which is always positive and smooth)—that high mode occupation alone is not always sufficient for such an interpretation and cannot be generally used to assert the 'classical'-like nature of the mode in question. The broadness of the Wigner

---

[1] We clarify our terminology here by noting that evolution of stochastic trajectories for a phase-space variable from some initial state (defined appropriately by a corresponding initial Wigner function) under a particular Hamiltonian, is completely equivalent to directly sampling this variable from the known Wigner function of the final state after said evolution.

distribution for the squeezed coherent states can, on the other hand, still serve as the sufficient condition.

This chapter is organized such that in Sect. 6.1 we demonstrate formally the underlying mathematical relation between P_n and $\tilde{P}_n$ in the Wigner representation and the conditions on $W_{|\psi\rangle}(\alpha)$ for $\tilde{P}_n$ to approximately correspond to P_n. In Sect. 6.2 we investigate quantitatively the legitimacy of the method by applying it to the thermal and squeezed coherent states. Finally, in Sect. 6.3 we examine under what conditions we expect the method to fail, and how such a failure would manifest in calculations.

6.1 Formal Derivation

To formally evaluate the particle number distribution P_n of a single-mode state $|\psi\rangle$, one may calculate the overlap of the state $|\psi\rangle$ with the Fock state $|n\rangle$, which in the Wigner representation is given by [3]

$$P_n \equiv |\langle\psi|n\rangle|^2 = \pi \int d^2\alpha W_{|\psi\rangle}(\alpha) W_{|n\rangle}(\alpha), \tag{6.1}$$

where $W_{|\psi\rangle}(\alpha)$ and $W_{|n\rangle}(\alpha)$ are the respective Wigner functions, with $W_{|n\rangle}(\alpha)$ given by [3]

$$W_{|n\rangle}(\alpha) = \frac{2}{\pi}(-1)^n e^{-2|\alpha|^2} L_n(4|\alpha|^2), \tag{6.2}$$

where $L_n(x)$ is the nth-order Laguerre polynomial. With knowledge of the explicit form of $W_{|\psi\rangle}(\alpha)$ one may then analytically or numerically evaluate the integral in Eq. (6.1) to derive the number distribution of the state exactly. In dynamical simulations one may numerically solve the integral (6.1) by first reconstructing the Wigner function $W_{|\psi\rangle}(\alpha)$ itself, or by noting that the RHS of Eq. (6.1) is formally equivalent to

$$P_n \equiv \pi \langle W_{|n\rangle}(\alpha) \rangle_W, \tag{6.3}$$

where the subscript refers to averaging over many stochastic trajectories which provide samples of α_i according to the distribution $W_{|\psi\rangle}(\alpha)$. Such a computation is in general non-trivial for highly occupied states or those with a sufficiently broad number distribution as it requires evaluation of high-order Laguerre polynomials with large arguments. Usually, computational techniques such as quadruple precision will be required to overcome numerical issues for $n \gtrsim 256$. Our analysis of $\tilde{P}_n$ thus has interest beyond the interpretation of individual stochastic trajectories of the Wigner function as the underlying binning formalism overcomes such computational issues, inherent to the exact method, and offers instead a much simpler method to implement numerically.

To characterize the connection of $\tilde{P}_n$ to this formal definition of P_n we can mathematically define the binned probability distribution as

$$\tilde{P}_n \equiv \int_n^{n+1} d(|\alpha|^2)\, \mathcal{P}(|\alpha|^2), \tag{6.4}$$

where $\mathcal{P}(|\alpha|^2)$ is the probability density of sampling $|\alpha|^2$ from an ensemble of stochastic trajectories. In terms of the Wigner function, this is equivalent to the probability of sampling α from within an annulus in phase-space with inner and outer radii of $\sqrt{n}$ and $\sqrt{n+1}$ respectively. Thus we may rewrite Eq. (6.4), using the Heaviside step function $\theta(x)$, as

$$\tilde{P}_n = \pi \int d^2\alpha \left[\frac{1}{\pi}\theta(|\alpha| - \sqrt{n})\theta(\sqrt{n+1} - |\alpha|) \right] W_{|\psi\rangle}(\alpha). \tag{6.5}$$

Comparing the result of Eq. (6.5) to Eq. (6.1) we see that the binning procedure is mathematically equivalent to replacing $W_{|n\rangle}(\alpha)$ by a radially symmetric boxcar function in phase-space defined as

$$\tilde{W}_{|n\rangle}(\alpha) = \frac{1}{\pi}\theta(|\alpha| - \sqrt{n})\theta(\sqrt{n+1} - |\alpha|). \tag{6.6}$$

This representation of the Fock state Wigner function is known as a Planck–Bohr–Sommerfeld band [4], and is equivalent to a smearing out of the classical (Kramers) trajectory of a Fock state in phase-space, which is a ring along $|\alpha| = \sqrt{n+1/2}$. The binning procedure as characterized by Eq. (6.5) is then similar to the area-of-overlap formalism developed previously by Schleich [4], wherein the number distribution of a state can be approximated by the overlap of the phase-space distribution with a band in phase-space, representing the number state. We point out the subtle difference that Schleich's formalism can account for interference between probability amplitudes, which is equivalent to retaining negative contributions in Eq. (6.1), whereas the binning procedure rules this out as Eq. (6.5) is a sum of contributions from a strictly non-negative Wigner function.

One can also justify the approximation of $\tilde{W}_{|n\rangle}(\alpha)$ by a more practical argument by noting that low-order moments of α with respect to $W_{|n\rangle}(\alpha)$ are dominated by contributions of the final 'crest' in the highly-oscillatory Wigner distribution (see Fig. 6.1 for illustration), whilst earlier contributions effectively cancel out. Such an approach is similar to the approximations applied by Gardiner et al. in Ref. [12], wherein the authors observed that the Wigner function of the Fock state could be approximated as a radially symmetric Gaussian ring,

$$W_{|n\rangle}(\alpha) = \mathcal{A}\, e^{-2(|\alpha|^2 - n - 1/2)^2}, \tag{6.7}$$

which is strictly positive ($\mathcal{A}$ being the normalization constant). In Refs. [29, 30] Olsen et al. demonstrated explicitly that sampling of $W_{|n\rangle}(\alpha)$ indeed produced all

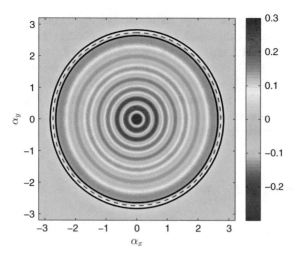

Fig. 6.1 Colormap plot of the Wigner distribution, $W_{|n\rangle}(\alpha)$ of the $n = 7$ Fock state, Eq. (6.2), where the axis correspond to $\alpha_x \equiv \text{Re}(\alpha)$ and $\alpha_y \equiv \text{Im}(\alpha)$. The radial oscillations appear distinctly as a series of alternating peaks ($W_{|n\rangle}(\alpha) > 0$) and troughs ($W_{|n\rangle}(\alpha) < 0$). For illustration, we overlay the Planck–Bohr–Sommerfeld band for the equivalent state, Eq. (6.6). The inner and outer radii (*solid lines*) are $\sqrt{n}$ and $\sqrt{n+1}$, which are centered around the 'classical' trajectory (*dashed line*) which is a ring of radius $\sqrt{n+1/2}$

moments $\langle |\alpha|^m \rangle_W$ of the exact Wigner distribution up to $\mathcal{O}(1/n^2)$ relative to the leading order, implying that the contribution of all but the final oscillation in $W_{|n\rangle}(\alpha)$ can be considered approximately negligible. In light of this, one could also regard $\tilde{W}_{|n\rangle}(\alpha)$, Eq. (6.6), as a further crude approximation to $\mathcal{W}_{|n\rangle}(\alpha)$.

Following the reasoning of Gardiner et al. [12], we thus intuitively expect the replacement of $W_{|n\rangle}(\alpha)$ by $\tilde{W}_{|n\rangle}(\alpha)$ in Eq. (6.5) to only be a good approximation when $W_{|\psi\rangle}(\alpha)$ is a sufficiently smooth function of α. Qualitatively, this means that we require $W_{|\psi\rangle}(\alpha)$ to be slowly varying on the order of the characteristic length scale of oscillations in $W_n(\alpha)$, which, using the properties of the Laguerre polynomial $L_n(4|\alpha|^2)$, can be estimated to be $\sim 1/\sqrt{n}$. There are two complementary properties of $W_{|\psi\rangle}(\alpha)$ which achieve this outcome. Firstly, for states localized near the origin in phase-space—such as the thermal state—one requires that the Wigner function has a characteristic width $\sigma \gg 1$. This implies that $\tilde{P}_n$ will approximate P_n well even for small $n \sim 1$. Secondly, for states of fixed width—such as the coherent or squeezed coherent states—one requires a large coherent displacement $|\beta|$ from the origin. As the overlap between $W_{|\psi\rangle}(\alpha)$ and $W_{|n\rangle}(\alpha)$ will generally be greatest for $n \sim |\beta|^2$, the length-scale of the oscillations in $W_{|n\rangle}(\alpha)$ in the relevant regions of $W_{|\psi\rangle}(\alpha)$ will scale as $1/|\beta|$. The width of $W_{|\psi\rangle}(\alpha)$ relative to the scale of these oscillations thus increases as $|\beta|$ increases, improving the validity of replacing $W_{|n\rangle}(\alpha)$ with $\tilde{W}_{|n\rangle}(\alpha)$. In the following section we illustrate these arguments both qualitatively and quantitatively for the thermal and squeezed coherent states.

Lastly, although this derivation has focused on the single-mode case it may be trivially generalized to a multi-mode state and an equivalent form of $\tilde{P}_{n_1,n_2,\ldots}$ may be found. The same generalized conditions regarding the relative width of the Wigner function may be applied. However, in the following section we will continue to focus our analysis on the single-mode case as it allows us to illustrate the correspondence between the two number distributions in a transparent manner. More specifically, we will use two particular states to analyse the similarity between P_n and $\tilde{P}_n$: (i) thermal and (ii) squeezed coherent states.

6.2 Similarity of P_n and $\tilde{P}_n$

6.2.1 Thermal State

The first state we consider is the thermal state, which is a mixed state defined by the density matrix

$$\hat{\rho}_{\text{th}} = \sum_{n=0}^{\infty} P_n |n\rangle \langle n|, \tag{6.8}$$

where the number distribution is given by [5]

$$P_n = \frac{\bar{n}^n}{(\bar{n}+1)^{n+1}}, \tag{6.9}$$

and is characterized solely by the mean occupation $\langle \hat{n} \rangle = \bar{n}$.

The corresponding Wigner function is [5]

$$W_{\text{th}}(\alpha) = \frac{1}{\pi(\bar{n}+1/2)} \exp\left(-\frac{|\alpha|^2}{\bar{n}+1/2}\right). \tag{6.10}$$

The rms width of this distribution is then $\sigma = \sqrt{(\bar{n}+1/2)/2}$ and, therefore, according to our criterion, the sufficient requirement ($\sigma \gg 1$) for $\tilde{P}_n$ to agree well with the physical P_n is equivalent in this case to high mean mode occupation $\bar{n} \gg 1$.

Substituting $W_{\text{th}}(\alpha)$ into Eq. (6.5) leads to

$$\tilde{P}_n = e^{-n/(\bar{n}+1/2)} \left[1 - e^{-n/(\bar{n}+1/2)}\right]. \tag{6.11}$$

Although this form of $\tilde{P}_n$ clearly differs from P_n, a keen eye will note that in fact

$$\tilde{P}_n = \frac{\langle n \rangle_{\text{bin}}^n}{(\langle n \rangle_{\text{bin}} + 1)^{n+1}}, \tag{6.12}$$

where

$$\langle n \rangle_{\text{bin}} \equiv \sum_{n=0}^{\infty} n \tilde{P}_n = \frac{1}{e^{1/(\bar{n}+1/2)} - 1}. \tag{6.13}$$

Hence while both distributions may be written solely in terms of their respective means, $\tilde{P}_n \neq P_n$ explicitly as $\langle n \rangle_{\text{bin}} \neq \bar{n}$.

As a quantitative measure of how well the binned particle number distribution $\tilde{P}_n$ approximates the true distribution P_n, we introduce the Bhattacharyya statistical distance, which is defined as [31]

$$D_B = -\ln[B(P, \tilde{P})], \tag{6.14}$$

where the Bhattacharyya coefficient is given by

$$B(P, \tilde{P}) = \sum_{n=0}^{\infty} \sqrt{P_n \tilde{P}_n}. \tag{6.15}$$

For $\tilde{P}_n \to P_n$ the Bhattacharyya coefficient becomes $B(P, \tilde{P}) \to \sum_{n=0}^{\infty} P_n = 1$ due to the normalization condition and hence $D_B \to 0$, indicating complete overlap of the distributions.

For the thermal state the Bhattacharyya coefficient can be calculated exactly to give

$$B(P, \tilde{P}) = \frac{\left[1 - e^{-2/(2\bar{n}+1)}\right]^{1/2}}{(\bar{n}+1)^{1/2} - \bar{n}^{1/2} e^{-1/(2\bar{n}+1)}}, \tag{6.16}$$

and thus the Bhattacharyya distance is

$$D_B = -\frac{1}{2} \ln \left[1 - e^{-2/(2\bar{n}+1)}\right] + \ln \left[\sqrt{\bar{n}+1} - \sqrt{\bar{n}} e^{-1/(2\bar{n}+1)}\right]. \tag{6.17}$$

In the limit of $\bar{n} \gg 1$ we find the behaviour

$$D_B \propto \bar{n}^{-4}, \tag{6.18}$$

which indicates that for large mean occupation $\tilde{P}_n$ rapidly approaches the true P_n. To illustrate this strong correspondence between P_n and $\tilde{P}_n$ we plot a comparison of the distributions for a thermal state with $\bar{n} = 10$ in Fig. 6.2a; as we see, even only moderately large mean occupations, such as in this example, render the two distributions visually identical, with a Bhattacharyya distance of $D_B = 6.63 \times 10^{-5}$.

We could recast the result of Eq. (6.18) in terms of the width of the distribution, $\sigma \simeq \sqrt{\bar{n}/2}$ for $\bar{n} \gg 1$, as

$$D_B \propto \sigma^{-8}. \tag{6.19}$$

Fig. 6.2 a Example of the true particle number distribution P_n (grey bars) for a thermal state with $\bar{n} = 10$, compared with the binned number distribution $\tilde{P}_n$ (red markers). **b** Statistical distance D_B between the two distributions, calculated from Eq. (6.17) for a range of mean occupations $\bar{n}$, which scales as $\propto 1/\bar{n}^4$

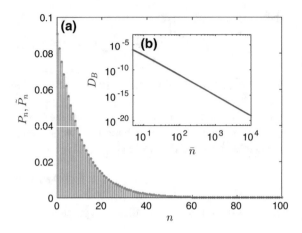

This strong scaling is a key result, particularly given that the statement of Blakie et al. (quoted in the Introduction section) pertains directly to the interpretation of c-field methods for Bose gases, for which this interpretation is applied directly to thermally populated states above the condensate mode. As we have shown, the heuristic link between Wigner trajectories and individual experimental runs, thought to be plausible for highly occupied states, can indeed be justified and quantified in terms of the similarity of $\tilde{P}_n$ and P_n. While for a thermal state, high mean occupation is actually equivalent to our sufficient requirement of having a broad Wigner distribution for this interpretation to be valid, there are situations (see next section) in which high mode occupation alone may not suffice for such an interpretation.

6.2.2   Squeezed Coherent State

The second state which we consider is the squeezed coherent state, defined as

$$|\beta, \eta\rangle = \hat{D}(\beta)\hat{S}(\eta)|0\rangle, \tag{6.20}$$

where $\hat{D}(\beta) = \exp(\beta\hat{a}^\dagger - \beta^*\hat{a})$ is the displacement operator and the squeezing operator is $\hat{S} = \exp[\{\eta^*\hat{a}^2 - \eta(\hat{a}^\dagger)^2\}/2]$ where $\eta = se^{i\theta}$ for $s \geq 0$ [5, 32]. In Fig. 6.3 we illustrate the actions of these operators in phase-space. Firstly the squeezing operator 'squeezes' the Gaussian Wigner distribution of the vacuum by an amount e^{-s} along an axis defined by the squeezing angle θ, whilst the perpendicular axis is stretched by e^s. The displacement operator then shifts the distribution in phase space by $\beta = |\beta|e^{i\varphi}$. There exist two special sub-cases of the squeezed coherent state: (i) the coherent state $|\beta\rangle$ where $\beta \neq 0$ and $s = 0$; and (ii) the squeezed vacuum state $|0, \eta\rangle$ where $\beta = 0$ and $s \neq 0$.

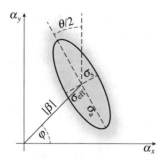

Fig. 6.3 Illustration of the Wigner function for a squeezed coherent state $W_{|\beta,\eta\rangle}(\alpha)$. The action of the squeezing operator $\hat{S}(\eta)$ on the initial state $|0\rangle$ is to squeeze the vacuum state Wigner function (a symmetric Gaussian with rms width $\sigma = 1/2$) by e^{-s} along the α_x-axis and stretch it by e^s along the α_y-axis, then rotate the distribution by $\theta/2$. The subsequent action of the displacement operator $\hat{D}(\beta)$ is to shift the distribution by $\beta = |\beta|e^{i\varphi}$. The relevant length scale in comparison to the radially-directed oscillations in $W_{|n\rangle}(\alpha)$ is the effective width σ_{eff} along the radial direction of $W_{|\beta,\eta\rangle}(\alpha)$

The Wigner function of the general squeezed coherent state can be written in a simple form [5]

$$W_{|\beta,\eta\rangle}(\gamma) = \frac{2}{\pi}\exp\left(-\frac{\gamma_x^2}{2\sigma_s^2} - \frac{\gamma_y^2}{2\sigma_a^2}\right), \tag{6.21}$$

where

$$\gamma_x = (\alpha_x - \beta_x)\cos\left(\frac{\theta}{2}\right) + (\alpha_y - \beta_y)\sin\left(\frac{\theta}{2}\right), \tag{6.22}$$

$$\gamma_y = -(\alpha_x - \beta_x)\sin\left(\frac{\theta}{2}\right) + (\alpha_y - \beta_y)\cos\left(\frac{\theta}{2}\right), \tag{6.23}$$

for $\alpha = \alpha_x + i\alpha_y$ and $\beta = \beta_x + i\beta_y$. The rms widths along the squeezed and anti-squeezed axes are given by $\sigma_s = e^{-s}/2$ and $\sigma_a = e^s/2$, respectively. Independent control over the parameters β and η allows us to quantitatively probe the similarity of $\tilde{P}_n$ and P_n as a function of the width of the Wigner distribution.

The number distribution of the squeezed state is nontrivial,

$$P_n = \frac{\left(\frac{1}{2}\tanh(s)\right)^n}{n!\cosh(s)}e^{-|\beta|^2[1+\cos(2\varphi-\theta)\tanh(s)]}$$

$$\times \left|H_n\left(\frac{\beta + \beta^* e^{i\theta}\tanh(s)}{\sqrt{2e^{i\theta}\tanh(s)}}\right)\right|^2. \tag{6.24}$$

with mean occupation $\langle \hat{n} \rangle = |\beta|^2 + \sinh^2(s)$ [32, 33]. For large coherent displacement such that $|\beta|^2 \gg e^{2s}$, this P_n can be approximated by a simple Gaussian [32]

$$P_n \simeq \frac{1}{\sqrt{2\pi\langle\Delta^2\hat{n}\rangle}} \exp\left[\frac{-(n - |\beta|^2)^2}{2\langle\Delta^2\hat{n}\rangle}\right], \tag{6.25}$$

whose rms width is given by $\sigma = \sqrt{\langle\Delta^2\hat{n}\rangle}$, where

$$\langle\Delta^2\hat{n}\rangle = |\beta|^2 \left[e^{-2s}\cos^2\left(\varphi - \frac{\theta}{2}\right) + e^{2s}\sin^2\left(\varphi - \frac{\theta}{2}\right)\right]. \tag{6.26}$$

This form demonstrates how the squeezing operator stretches or squeezes the probability distribution P_n according to the relative orientation of the squeezing and coherent displacement. In this section, our analysis will be limited to a range of squeezing such that the above approximation for P_n is valid. The effects of stronger squeezing and its implications for both P_n and $\tilde{P}_n$ will be discussed in Sect. 6.3.

An analytic form of $\tilde{P}_n$ can be found by substituting Eq. (6.21) into the definition of Eq. (6.5), however, the result is not particularly insightful. We point the interested reader to Ref. [34] as a guide to the general form of the calculation. Instead, we numerically evaluate $\tilde{P}_n$ by stochastically sampling $W_{|\beta,\eta\rangle}(\alpha)$ according to the prescription of Ref. [29] and binning the calculated occupation of each sample. Such a construction is equivalent to obtaining the same state and results via a dynamical simulation of stochastic equations (trajectories) in the Wigner representation, as the phenomenological squeezed vacuum state can be generated from a Hamiltonian for spontaneous parametric down-conversion (in the undepleted pump approximation) $\hat{H} = i\hbar[g^*\hat{a}^2 - g(\hat{a}^\dagger)^2]$, in which case the squeezing parameter η is actually given by $\eta \equiv gt$. The subsequent coherent displacement of the squeezed state is achieved by coupling the mode $\hat{a}$ to a classical field of amplitude ε, equivalent to evolution under the Hamiltonian $\hat{H} = i\hbar\kappa[\varepsilon^*\hat{a} - \varepsilon\hat{a}^\dagger]$ where κ is the coupling strength and hence the resulting displacement is related as $\beta \equiv \kappa\varepsilon t$.

We numerically evaluate the Bhattacharyya distance as a function of coherent displacement for some example squeezed coherent states with $\varphi = 0$, $s = 0.4$ and squeezing angles of $\theta = 0$ and $\theta = \pi$, which are referred to as amplitude- and phase-squeezing respectively. The results are plotted in the inset to Fig. 6.4. Also plotted is the simple case of the coherent state for which $s = 0$, whilst other parameters are kept identical. We find a generic scaling independent of s,

$$D_B \propto |\beta|^{-2}, \tag{6.27}$$

in the regime where $|\beta|^2 \gg e^{2s}$ and the approximate form of Eq. (6.25) is valid. This result predicts a rapid convergence of $\tilde{P}_n$ to P_n with increasing occupation $\langle\hat{n}\rangle \approx |\beta|^2$. For $|\beta|^2 = 50$ and for the three cases of squeezing, the calculated distributions $\tilde{P}_n$ and P_n are plotted in the main panel of Fig. 6.4 and are visually indistinguishable from each other.

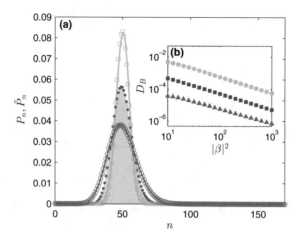

Fig. 6.4 **a** Example of probability distribution $\tilde{P}_n$ (*red markers*) for a coherent state with $|\beta|^2 = 50$, compared to P_n (*grey bars*). Also plotted are the distributions $\tilde{P}_n$ (*markers*) P_n (*lines*) and for a squeezed coherent state with $s = 0.4$, $\theta = 0$ (*magenta circles*) and $\theta = \pi$ (*green squares*). Excellent qualitative agreement is found between the two distributions in all cases. **b** Quantitative comparison of $\tilde{P}_n$ and P_n by the statistical distance D_B for a coherent state (*blue squares*), squeezed state with $s = 0.2$ and $\theta = 0$ (*green circles*) and squeezed state with $s = 0.2$ and $\theta = \pi$ (*magenta triangles*). We find a consistent scaling of $D_B \propto 1/|\beta|^2$ for all three cases. Stochastic sampling error of one standard deviation is not indicated but is less than 2 % of calculated D_B for all data points (obtained from approximately 10^9 trajectories)

Beyond the scaling with coherent displacement, we may also examine how the absolute width of the Wigner function affects the agreement of $\tilde{P}_n$ with P_n by manipulation of the squeezing strength s and angle θ. The relevant length scale will be the effective width σ_{eff} of the distribution (see Fig. 6.3) with respect to the radially directed oscillations in $W_{|n\rangle}(\alpha)$,

$$\sigma_{\text{eff}} = \sqrt{\sigma_s^2 \cos^2\left(\varphi - \frac{\theta}{2}\right) + \sigma_a^2 \sin^2\left(\varphi - \frac{\theta}{2}\right)}. \tag{6.28}$$

We plot the dependence of the Bhattacharyya distance as a function of this parameter in Fig. 6.5a and find it scales as

$$D_B \propto \sigma_{\text{eff}}^{-6}, \tag{6.29}$$

independently of coherent displacement β. This strong scaling again agrees with our intuitive argument, indicating that $\tilde{P}_n$ rapidly approaches P_n as the Wigner function becomes increasingly smooth on the length scale of oscillations in $W_{|n\rangle}(\alpha)$. We note the difference to the scaling of the thermal state is partly attributable to the difference that the squeezed state is a minimum uncertainty state, meaning that an increase in the width of one axis (σ_a) is offset by a decrease in the perpendicular axis (σ_s) such

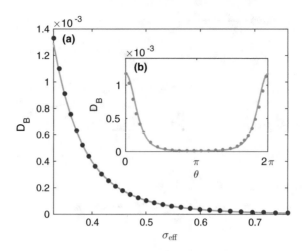

Fig. 6.5 **a** Behaviour of statistical distance D_B with the effective width σ_{eff} for a squeezed coherent state with $|\beta|^2 = 50$ and $\varphi = 0$. For $\sigma_{eff} \leq 1/2$ we calculate D_B by fixing the squeezing angle as $\theta = 0$ and thus $\sigma_{eff} \equiv \sigma_s \leq 1/2$. Alternately, for $\sigma_{eff} \geq 1/2$ we fix the squeezing angle as $\theta = \pi$ and thus $\sigma_{eff} \equiv \sigma_a \geq 1/2$. A fit $D_B \propto \sigma_{eff}^{-6}$ (*grey line*) is also plotted for comparison with the actual stochastically sampled data (*blue circles*). **b** Variation of D_B with squeezing angle θ for a squeezed coherent state with $|\beta|^2 = 50$, $\varphi = 0$ and $s = 0.4$ (*green circles*). The behaviour fits the model of Eq. (6.29) (*grey line*) where σ_{eff} depends on the squeezing angle θ as per Eq. (6.28). For numerically calculated data in both **a** and **b** stochastic sampling error of one standard deviation is less than 2 % of calculated value (obtained from approximately 10^9 trajectories)

that $\sigma_a \sigma_s = 1/4$ is preserved. This is in contrast to the thermal state which has a radially symmetric rms width which increases with average occupation.

In Fig. 6.5b we also plot the Bhattacharyya distance as a function of the squeezing angle. For a state with a purely real coherent displacement ($\varphi = 0$), we find D_B is minimal for phase-squeezed states ($\theta = \pi$) and maximal for amplitude squeezed states ($\theta = 0$). For phase-squeezing, the anti-squeezed axis of the distribution is aligned radially, along the direction of the oscillations in $W_{|n\rangle}(\alpha)$, and σ_{eff} is maximal. We thus expect for this scenario that our approximation $\tilde{W}_{|n\rangle}(\alpha)$ should be the most valid as any oscillations will be averaged out in Eq. (6.1), leading to minimal D_B. Conversely, for amplitude squeezing the squeezed axis of the distribution is aligned radially, minimizing σ_{eff} and thus we expect our approximation to be the least valid, leading to a larger D_B.

6.3 Breakdown of Relationship

The analysis of the previous section has demonstrated how in general, $\tilde{P}_n$ closely replicates P_n when the relative width of the Wigner distribution $W_{|\psi\rangle}(\alpha)$ is large compared to the oscillation period of the Fock state Wigner function, $W_{|n\rangle}(\alpha)$.

However, one can also find a few simple counter-examples to demonstrate how the correspondence breaks down when the underlying approximations are no longer valid. In particular we demonstrate this with states that are highly-occupied, showing that large occupation alone is not sufficient for approximating P_n by $\tilde{P}_n$.

As an example, in Fig. 6.6 we plot P_n and $\tilde{P}_n$ for $|\beta|^2 = 20$, $s = 1.5$ and for two squeezing angles: (a) $\theta = 0$ and (b) $\theta = \pi$. In both cases we see a region emerges wherein the probability of odd and even n oscillates strongly. For amplitude squeezing ($\theta = 0$) these oscillations arise for $n \gtrsim |\beta|^2$ as predicted by Schleich and Wheeler [35]. In terms of the binning procedure it is clear that $W_{|\psi\rangle}(\alpha)$ is sufficiently elongated that it is approximately the width of the oscillations in $W_{|n\rangle}(\alpha)$ and multiple oscillations become important in the calculation of Eq. (6.1) as illustrated in Fig. 6.6b. Similar arguments apply to the case of phase squeezing ($\theta = \pi$), illustrated in Fig. 6.6c, d. In both cases, the narrowness of the Wigner distribution implies it is not valid to approximate $W_{|n\rangle}(\alpha)$ with $\tilde{W}_{|n\rangle}(\alpha)$ and thus $\tilde{P}_n$ does not well approximate P_n.

Related issues arise for the squeezed vacuum state, which can be considered an extreme case of the above examples wherein $|\beta|^2 = 0$ and the Wigner distribution is centered at the phase-space origin. The state is notable for its even-odd oscillatory number distribution,

$$P_{2m} = \frac{[\tanh(s)/2]^{2m}}{(2m)!\cosh(s)}|H_{2m}(0)|^2, \quad P_{2m+1} = 0. \tag{6.30}$$

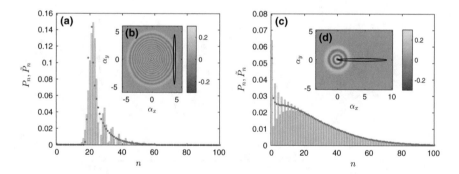

Fig. 6.6 a Probability distribution for a squeezed coherent state with $s = 1.5$, $\theta = 0$ and $|\beta|^2 = 20$. For $n \gtrsim |\beta|^2$ the true distribution P_n (*grey bars*) displays oscillations which are not replicated by $\tilde{P}_n$ (*red markers*). We construct $\tilde{P}_n$ from $\sim 10^7$ trajectories and stochastic sampling error is negligibly small. **b** The Wigner function of the $n = 25$ Fock state overlaid with an ellipse representing the $2\sigma_{a,s}$ contour of $W_{|\beta,\eta\rangle}(\alpha)$. This illustrates how the oscillations of $W_{|n\rangle}(\alpha)$, which we ignore in calculation of $\tilde{P}_n$, play an important role in calculating the overlap integral [Eq. (6.1)] for $n \gtrsim |\beta|^2$. **c** Same as **a** except $\theta = \pi$. For this squeezing angle we see that oscillations in P_n (*grey bars*) emerge for small n and are again not replicated in $\tilde{P}_n$ (*red markers*). **d** The Wigner function of the $n = 5$ Fock state, again overlaid with a $2\sigma_{a,s}$ contour of $W_{|\beta,\eta\rangle}(\alpha)$. The central dip of the Wigner function in both **b** and **d** $[W_{|n\rangle}(0) = -2/\pi$ for odd $n]$ saturates the colormap so as to allow better illustration of the oscillations

Again, in terms of phase-space representation of the state, this property is an effect of the narrowness of $W_{|0,\eta\rangle}(\alpha)$ combined with the negativity of the true Fock state Wigner function $W_{|n\rangle}(\alpha)$. Setting $\theta = 0$ for definitiveness, the Wigner distribution has an rms width of $\sigma_s \leq 1/2$ in the α_x direction, and thus it will obviously be sufficiently narrow to probe the individual oscillations of $W_{|n\rangle}(\alpha)$, which have a period on the order of 1 for small n. Accordingly, the interpretation of $\tilde{P}_n \sim P_n$ is not valid as the replacement of $W_{|n\rangle}(\alpha)$ by $\tilde{W}_{|n\rangle}(\alpha)$ in Eq. (6.1) is a poor approximation.

In Fig. 6.7 we plot the Bhattacharyya distance for a broader range of squeezed coherent states, highlighting specifically the regimes in which $\tilde{P}_n$ replicates P_n and where this breaks down. We find for $\theta = \pi$ ($\sigma_{\text{eff}} \geq 1/2$) the Bhattacharyya distance behaves according to the power-law of Eq. (6.29) (indicated by the linear regime on the log-log axes) until a turning point $\sigma_{\text{eff}} \approx 0.84|\beta|^{2/3}$, where the oscillations of $W_{|n\rangle}(\alpha)$ near the origin become important for small n. For $\theta = 0$ ($\sigma_{\text{eff}} \leq 1/2$) the statistical distance worsens due to the narrowness of the distribution according to Eq. (6.29) (again, the linear regime) until the emergence of oscillations. The transition from the linear relationship occurs in the vicinity $\sigma_{\text{eff}} \approx 1/(2|\beta|^{1/3})$, which agrees with that regarding the emergence of oscillations in P_n as previously studied in Ref. [35].

When we examine the forementioned states for which our procedure does not reproduce P_n accurately, we see that these are states for which the quantization of

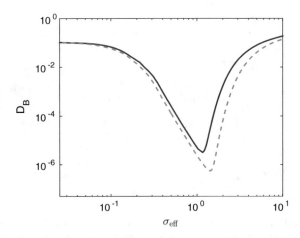

Fig. 6.7 Dependence of statistical distance D_B on the effective width σ_{eff} of the squeezed coherent state Wigner distribution for $|\beta|^2 = 20$ (*blue solid line*) and $|\beta|^2 = 40$ (*red dashed line*). Without loss of generality we arbitrarily set $\varphi = 0$ for all states. Identically to Fig. 6.5a we calculate $\sigma_{\text{eff}} \leq 1/2$ by setting $\theta = 0$ and thus $\sigma_{\text{eff}} \equiv \sigma_s$, and similarly $\sigma_{\text{eff}} > 1/2$ by $\theta = \pi$ and $\sigma_{\text{eff}} \equiv \sigma_a$. Stochastic sampling error of one standard deviation is not indicated, however, it is restricted to less than 2 % of calculated D_B values. For relatively weak squeezing ($|\beta|^2 \gg e^{2s}$), the power-law scaling of Fig. 6.5a is illustrated by the linear-regime ($0.4 \lesssim \sigma_{\text{eff}} \lesssim 1$) with the logarithmic scale. The deviation from the linear regime, indicating oscillatory structure in P_n which is not replicated by $\tilde{P}_n$, occurs at $\sigma_{\text{eff}} \approx 1/(2|\beta|^{1/3})$ for $\theta = 0$, whilst there is an obvious turning point in D_B at $\sigma_{\text{eff}} \approx 0.84|\beta|^{2/3}$ for $\theta = \pi$

the field is important. The squeezed vacuum is a prime example of this, containing only even numbers of photons. As we coherently displace this state from the origin (squeezed coherent state), the displacement becomes more important than the squeezing and $\tilde{P}_n$ becomes more accurate. This is consistent with the fact that the truncated or a priori positive Wigner distribution is often described as equivalent to the classical theory of stochastic electrodynamics [36].

6.4 Conclusion

In summary, we have examined under which conditions a naive calculation of the binned number distribution from individual (truncated) Wigner trajectories, $\tilde{P}_n$, can replicate closely the true particle number distribution P_n, hence justifying the interpretation of these trajectories as representing individual experimental outcomes. The sufficient requirement for this is that the Wigner function $W_{|\psi\rangle}(\alpha)$ of the state $|\psi\rangle$ varies sufficiently smoothly on the characteristic length scale of oscillations in the Wigner function $W_{|n\rangle}(\alpha)$ of the Fock state $|n\rangle$. This is, of course, in addition to the constraint that only positive Wigner functions $W_{|\psi\rangle}(\alpha)$ are being considered, which is the case in the truncated Wigner approximation or in model Hamiltonians that depend no-higher-than quadratically on creation or annihilation operators.

We have provided a rigorous operational definition of this seemingly heuristic binning procedure as one that corresponds to approximating the Wigner function of the Fock state (which appears in the definition of P_n via an overlap integral with the Wigner function of the state of interest) as a boxcar function in phase space. For states localized around the phase-space origin (e.g., a thermal state), the requirement of smoothness of the Wigner function is satisfied by a broad distribution, having a characteristic width much larger than unity. In this case, the large width of the distribution is equivalent to having large mode occupation number. On the other hand, for states that have large coherent displacement β (such as coherent and squeezed coherent states with $|\beta| \gg 1$), one can tolerate relatively narrow Wigner functions as long as its width remains much larger than $1/|\beta|$, which is the characteristic length scale of oscillations in $W_{|n\rangle}(\alpha)$ for the most relevant values of n ($\sim|\beta|^2$). This condition is satisfied for coherent states and weakly squeezed states, but will break down for highly squeezed states when the width of the respective Wigner function in the narrow dimension becomes comparable to $1/|\beta|$, even though the mode occupation for such states can be very high. The latter case serves as a counterexample to the view that high mode occupation alone is sufficient to interpret individual Wigner trajectories as 'samples' of single experimental runs.

Although we have considered only a small subset of Wigner functions in this article, we expect that the relationship between $\tilde{P}_n$ and P_n established on an analysis of the width and displacement of the Wigner function will allow an easy application to further states. Importantly, in the truncated Winger formalism the reconstruction of an *a priori* unknown single-mode Wigner function from many stochastic trajectories is relatively trivial and allows one to extract the characteristic length scale of the quasidistribution and thus, according to our criterion, justify or reject the approximation $\tilde{P}_n$ with no knowledge of the exact P_n.

References

1. Wigner, E.: On the quantum correction for thermodynamic equilibrium. Phys. Rev. **40**, 749–759 (1932)
2. Moyal, J.E.: Quantum Mechanics as a Statistical Theory. Mathematical Proceedings of the Cambridge Philosophical Society, vol. 45, pp. 99–124. Cambridge University Press, Cambridge (1949)
3. Leonhardt, U.: Essential Quantum Optics. Cambridge University Press, Cambridge (2010)
4. Schleich, W.P.: Quantum Optics in Phase Space. Wiley, New York (2011)
5. Walls, D.F., Milburn, G.: Quantum Optics. Springer Study Edition. Springer, New York (1995)
6. Drummond, P.D., Hardman, A.D.: Simulation of quantum effects in Raman-active waveguides. EPL (Europhy. Lett.) **21**, 279 (1993)
7. Werner, M.J., Raymer, M.G., Beck, M., Drummond, P.D.: Ultrashort pulsed squeezing by optical parametric amplification. Phys. Rev. A **52**, 4202–4213 (1995)
8. Steel, M.J., et al.: Dynamical quantum noise in trapped Bose-Einstein condensates. Phys. Rev. A **58**, 4824 (1998)
9. Sinatra, A., Castin, Y., Lobo, C.: A Monte Carlo formulation of the Bogoliubov theory. J. Mod. Opt. **47**, 2629 (2000)
10. Sinatra, A., Lobo, C., Castin, Y.: Classical-field method for time dependent Bose-Einstein condensed gases. Phys. Rev. Lett. **87**, 210404 (2001)
11. Sinatra, A., Lobo, C., Castin, Y.: The truncated Wigner method for Bose-condensed gases: limits of validity and applications. J. Phys. B **35**, 3599 (2002)
12. Gardiner, C.W., Anglin, J.R., Fudge, T.I.A.: The stochastic Gross-Pitaevskii equation. J. Phys. B **35**, 1555 (2002)
13. Polkovnikov, A.: Evolution of the macroscopically entangled states in optical lattices. Phys. Rev. A **68**, 033609 (2003)
14. Norrie, A.A., Ballagh, R.J., Gardiner, C.W.: Quantum turbulence in condensate collisions: an application of the classical field method. Phys. Rev. Lett. **94**, 040401 (2005)
15. Norrie, A.A., Ballagh, R.J., Gardiner, C.W.: Quantum turbulence and correlations in Bose-Einstein condensate collisions. Phys. Rev. A **73**, 043617 (2006)
16. Ruostekoski, J., Isella, L.: Dissipative quantum dynamics of bosonic atoms in a shallow 1d optical lattice. Phys. Rev. Lett. **95**, 110403 (2005)
17. Isella, L., Ruostekoski, J.: Nonadiabatic dynamics of a Bose-Einstein condensate in an optical lattice. Phys. Rev. A **72**, 011601 (2005)
18. Isella, L., Ruostekoski, J.: Quantum dynamics in splitting a harmonically trapped bose-einstein condensate by an optical lattice: truncated wigner approximation. Phys. Rev. A **74**, 063625 (2006)
19. Deuar, P., Drummond, P.D.: Correlations in a BEC collision: first-principles quantum dynamics with 150 000 atoms. Phys. Rev. Lett. **98**, 120402 (2007)
20. Polkovnikov, A.: Phase space representation of quantum dynamics. Ann. Phys. **325**, 1790–1852 (2010)
21. Corney, J.F., Olsen, M.K.: Non-Gaussian pure states and positive wigner functions. Phys. Rev. A **91**, 023824 (2015)
22. Hudson, R.L.: When is the Wigner quasi-probability density non-negative? Rep. Math. Phys. **6**, 249–252 (1974)
23. Blakie, P.B., Bradley, A.S., Davis, M.J., Ballagh, R.J., Gardiner, C.W.: Dynamics and statistical mechanics of ultra-cold Bose gases using c-field techniques. Adv. Phys. **57**, 363–455 (2008)

24. Martin, A.D., Ruostekoski, J.: Quantum and thermal effects of dark solitons in a one-dimensional Bose gas. Phys. Rev. Lett. **104**, 194102 (2010)
25. Witkowska, E., Deuar, P., Gajda, M., Rzazewski, K.: Solitons as the early stage of quasicondensate formation during evaporative cooling. Phys. Rev. Lett. **106**, 135301 (2011)
26. Karpiuk, T., et al.: Spontaneous solitons in the thermal equilibrium of a quasi-1d Bose gas. Phys. Rev. Lett. **109**, 205302 (2012)
27. Javanainen, J., Ruostekoski, J.: Emergent classicality in continuous quantum measurements. New J. Phys. **15**, 013005 (2013)
28. Lee, M.D., Ruostekoski, J.: Classical stochastic measurement trajectories: Bosonic atomic gases in an optical cavity and quantum measurement backaction. Phys. Rev. A **90**, 023628 (2014)
29. Olsen, M.K., Bradley, A.S.: Numerical representation of quantum states in the positive-P and Wigner representations. Opt. Commun. **282**, 3924–3929 (2009)
30. Olsen, M.K., Bradley, A.S., Cavalcanti, S.B.: Fock-state dynamics in Raman photoassociation of Bose-Einstein condensates. Phys. Rev. A **70**, 033611 (2004)
31. Bhattacharyya, A.: On a measure of divergence between two statistical populations defined by their probability distributions. Bull. Calcutta Math. Soc. **35**, 99–109 (1943)
32. Loudon, R., Knight, P.L.: Squeezed light. J. Mod. Opt. **34**, 709–759 (1987)
33. Yuen, H.P.: Two-photon coherent states of the radiation field. Phys. Rev. A **13**, 2226–2243 (1976)
34. Gilliland, D.C.: Integral of the bivariate normal distribution over an offset circle. J. Am. Stat. Assoc. **57**, 758–768 (1962)
35. Schleich, W., Wheeler, J.A.: Oscillations in photon distribution of squeezed states and interference in phase space. Nature **326**, 574–577 (1987)
36. Marshall, T.W.: Random electrodynamics. Proc. R. Soc. Lond. Ser. A. Math. Phys. Sci. **276**, 475–491 (1963)

Chapter 7
Conclusion

In this thesis, we have investigated how non-classical correlations and entanglement between massive particles can be generated, characterized and measured in systems of ultracold atomic gases. In particular, we have demonstrated how, in the simplest approximation, the processes of spontaneous four-wave mixing in colliding BECs and spin-changing collisions in spinor condensates produce the archetypal two-mode squeezed vacuum state, which is known to show these features. Specifically, we have outlined theoretical proposals to demonstrate non-classical correlations, EPR entanglement and quantum nonlocality (in the sense of a violation of a Bell inequality) in these systems. Our analysis includes the construction of new and appropriate measurement protocols to identify and quantify these phenomena, whilst also identifying realistic parameter regimes for their experimental demonstration. Furthermore, our detailed theoretical treatment using stochastic numerical simulations has incorporated technical and experimental effects to investigate the robustness of these effects to practical and fundamental limitations.

In Chap. 2 we have shown that an atom-optics analog of the Hong–Ou–Mandel effect can be realized utilizing pair-correlated atoms produced via colliding Bose-Einstein condensates and laser-induced Bragg pulses (which are the atom-optics equivalent to optical mirrors and beam-splitters). By defining an appropriate measurement protocol which takes into account the multimode nature of the scattering halo, we have predicted a HOM dip visibility of $\simeq 69\%$ which indicates that the atom-atom correlations produced by the collision process are stronger than classically allowed. The first atomic HOM effect in a closely related setup of four-wave mixing in an optical lattice potential was subsequently realized in the metastable Helium lab in Palaiseau [1]. Whilst such a result is a pre-requisite for more fundamental tests of quantum mechanics, such as violation of a Bell inequality, the interferometric scheme itself serves as an important stepping stone for experimental demonstrations of such tests.

We have also demonstrated, in Chap. 3, that the same process of spontaneous four-wave mixing in colliding condensates can be utilized as a source of pair-correlated atoms for a violation of a motional-state Bell inequality in an atom-optics analog of the Rarity–Tapster interferometer. Our numerical simulations predict a violation of

© Springer International Publishing Switzerland 2016
R.J. Lewis-Swan, *Ultracold Atoms for Foundational Tests of Quantum Mechanics*, Springer Theses, DOI 10.1007/978-3-319-41048-7_7

the CHSH-Bell inequality ($S > 2$) for a range of parameters well within currently accessible experimental regimes, in reasonable agreement with the scaling of simple toy models based on idealized two-mode squeezed vacuum states. These results take into account physically important processes such as: the multimode nature of the halo, spatial evolution and expansion of the source condensates and the consequential effects on the pair-production process, and phase dispersion. Furthermore, we fully model the real-time application of the Bragg pulses, which explicitly allows for experimental imperfections such as loss into higher-order Bragg scattering modes and deviations from the ideal beam-splitter and mirror models. All of these effects are non-existent or negligible in analogous photonic experiments and not captured by simpler toy models of the scheme. Such a detailed analysis is crucial to guide future experiments, not only of this specific system but also related proposals [2–4].

In Chap. 4 we have demonstrated that spin-changing collisions in a spinor BEC are an ideal candidate to realize and verify EPR entanglement between massive particles. Specifically, focusing on the recent experiment of Ref. [5], we have theoretically demonstrated that when spin-changing collisions are initiated via vacuum fluctuations, a strong suppression of the EPR criterion can be achieved. However, we demonstrate that sources of noise, such as initial thermal fluctuations (which are usually neglected in analogous optical systems), play a crucial role in the viability of the atom-optics scheme. Our calculations have shown that a small thermal seed ($\simeq 1$ atom) initially in the $m_F = \pm 1$ substates is sufficient to destroy EPR entanglement in the system considered in Ref. [5] (corresponding to an initial condensate of 150–200 atoms). Complementary to this, we have also demonstrated that a more relaxed criteria of entanglement, in terms of inseparability, is far less sensitive to the form of the fluctuations which initiate the spin-changing collisions.

In Chap. 5, we have discussed an application of the two-mode squeezed vacuum state in the context of quantum metrology. Specifically, we have demonstrated how the pair-production process of spin-changing collisions in a spinor condensate is an excellent candidate to realize an atomic $SU(1, 1)$ interferometer, which in the undepleted pump approximation has an interferometric sensitivity at the Heisenberg limit. Our analysis has also highlighted a key difference between the atomic and photonic realizations of the interferometer. Specifically, in the atom-optics scheme the $m_F = \pm 1$ states must be sufficiently detuned from the pump mode (or isolated by transferal between hyperfine levels) such that the spin-changing collisions are completely halted and the archetypal $SU(1, 1)$ scheme is properly realized. In contrast, in an optical realization utilizing spontaneous parametric down-conversion, this control is trivially realized by the finite length of the $\chi^{(2)}$ nonlinear medium. It is hoped that this theoretical understanding of the atom-optics system will lead to an experimental realization of an atomic $SU(1, 1)$ interferometer in the near future [6].

Finally, in Chap. 6 we have examined under what conditions one may associate individual stochastic trajectories of the Wigner representation with the outcomes of individual experimental realizations. In particular, we have demonstrated that the binned number distribution from individual stochastic Wigner trajectories, $\tilde{P}_n$, can replicate closely the true particle number distribution P_n, a neccesary condition for the interpretation of these trajectories as representing individual experimental outcomes. By giving a rigorous definition of $\tilde{P}_n$ we have found that a sufficient requirement for this correspondence is that the Wigner function $W_{|\psi\rangle}(\alpha)$ of the state $|\psi\rangle$ is strictly non-negative and varies sufficiently smoothly on the characteristic length scale of oscillations in the Wigner function $W_{|n\rangle}(\alpha)$ of the Fock state $|n\rangle$. In general this conditions leads to the consequence that the state is highly occupied, agreeing with the broadly accepted, but heuristic, view that, for highly occupied states, individual stochastic trajectories of the Wigner function correspond to outcomes of single experiments. However, we have also shown a counterexample, wherein the correspondence between $\tilde{P}_n$ and P_n breaks down for strongly squeezed states when the width of the respective Wigner function in the narrow dimension becomes comparable to the length scale of oscillations in $W_{|n\rangle}(\alpha)$, even though the state may be highly occupied.

References

1. Lopes, R., et al.: Atomic Hong-Ou-Mandel experiment. Nature **520**, 66 (2015)
2. Bonneau, M., et al.: Tunable source of correlated atom beams. Phys. Rev. A **87**, 061603 (2013)
3. Kheruntsyan, K.V., Olsen, M.K., Drummond, P.D.: Einstein-Podolsky-Rosen correlations via dissociation of a molecular Bose-Einstein condensate. Phys. Rev. Lett. **95**, 150405 (2005)
4. Bücker, R., et al.: Twin-atom beams. Nat. Phys. **7**, 608 (2011)
5. Gross, C., et al.: Atomic homodyne detection of continuous-variable entangled twin-atom states. Nature **480**, 219 (2011)
6. Linnemann, D., Lewis-Swan, R.J., Strobel, H., Mussel, W., Kheruntstyan, K.V., Oberthaler, M.K.: Quantum-enhanced sensing based on time reversal of non-linear dynamics (2016). arXiv:1602.07505

Appendix A
Analytic Models of Condensate Collisions

The dynamics of spontaneous four-wave mixing has been the subject of intense theoretical study in recent years. Various analytic and numerical techniques have been used, focused on the production of scattered atom pairs [1–4]. We present here two analytic techniques which, while subject to constraints in terms of physical applicability, provide an excellent starting point in terms of understanding the origin of various phenomena which arise through the collision process. Both techniques have been presented previously in the literature [1, 3], in particular we point readers to Ref. [3] for further details regarding the perturbative technique presented below.

A.1 Homogeneous Condensates in the Undepleted Pump Approximation

The first analytic technique we consider is the simplest 'toy model' of the four-wave mixing process. Although it may appear crude, it has proven invaluable in simple analysis of condensate collisions and encapsulates many physical features of the full process.

An effective Hamiltonian of the scattering process, in the Bogoliubov approximation, can be written in the form [5, 6]

$$
\begin{aligned}
\hat{H}_{\text{eff}} = \int d^3\mathbf{r} \, \Bigg\{ & \hat{\delta}^\dagger(\mathbf{r}, t) \left[-\frac{\hbar^2}{2m} \nabla^2 \right] \hat{\delta}(\mathbf{r}, t) + 2U \, |\psi(\mathbf{r}, t)|^2 \, \hat{\delta}^\dagger(\mathbf{r}, t) \hat{\delta}(\mathbf{r}, t) \\
& + U \left[\psi_{+\mathbf{k}_0}(\mathbf{r}, t) \psi_{-\mathbf{k}_0}(\mathbf{r}, t) \hat{\delta}^\dagger(\mathbf{r}, t) \hat{\delta}^\dagger(\mathbf{r}, t) \right. \\
& \left. + \psi^*_{+\mathbf{k}_0}(\mathbf{r}, t) \psi^*_{-\mathbf{k}_0}(\mathbf{r}, t) \hat{\delta}(\mathbf{r}, t) \hat{\delta}(\mathbf{r}, t) \right] \Bigg\},
\end{aligned}
\tag{A.1}
$$

where $U = 4\hbar^2 a/m$ is the interaction strength between atoms in the condensate and a is the s-wave scattering length. To derive this Hamiltonian we implemented

© Springer International Publishing Switzerland 2016
R.J. Lewis-Swan, *Ultracold Atoms for Foundational Tests of Quantum Mechanics*, Springer Theses, DOI 10.1007/978-3-319-41048-7

a Bogoliubov approximation of the wavefunction, wherein the full bosonic field operator $\hat{\psi}$ is split into mean-field and fluctuating components

$$\hat{\psi}(\mathbf{r}, t) = \psi_{+\mathbf{k}_0}(\mathbf{r}, t) + \psi_{-\mathbf{k}_0}(\mathbf{r}, t) + \hat{\delta}(\mathbf{r}, t). \tag{A.2}$$

The fluctuating component $\hat{\delta}$ represents the scattered atoms which populate the collision halo and is treated to lowest order in perturbation theory. The initially split condensate is represented by the two counter-propagating mean-field components, $\psi_{+\mathbf{k}_0}$ and $\psi_{-\mathbf{k}_0}$ (with momenta $\pm\mathbf{k}_0$ respectively). In this model we assume for simplicity that the condensate has a uniform density profile in position space and it is treated in the undepleted pump approximation. The wavefunctions can then be written as

$$\psi_{\pm\mathbf{k}_0}(\mathbf{r}, t) = \sqrt{\frac{\rho_0}{2}}\exp\left(\pm i\mathbf{k}_0 \cdot \mathbf{r} - \frac{i\hbar k_0}{2m}t\right), \tag{A.3}$$

where ρ_0 is the uniform density of the condensate and for simplicity we herein define $k_0 = |\mathbf{k}_0|$. The undepleted pump approximation is only valid for short times, which in general corresponds to ensuring that the occupation of the collision halo does not exceed 10 % of the initial condensate population.

Neglecting the effective mean-field potential felt by the scattered atoms due to the condensates, which is justified when the kinetic energy of the scattered atoms $\hbar k_0^2/2m$ is much higher than the mean-field interaction energy per particle in the condensate, the Heisenberg equation of motion for the fluctuating field is given by

$$\frac{\partial\hat{\delta}(\mathbf{r}, t)}{\partial t} = -\frac{i\hbar}{2m}\nabla^2\hat{\delta}(\mathbf{r}, t) - i\frac{U}{\hbar}\rho_0\exp\left(-\frac{i\hbar k_0}{m}t\right)\hat{\delta}^\dagger(\mathbf{r}, t). \tag{A.4}$$

Transforming to a rotating frame $\hat{\bar{\delta}}(\mathbf{r}, t) = \hat{\delta}(\mathbf{r}, t)\exp(i\hbar k_0^2 t/2m)$ and introducing the Fourier transform pair

$$\hat{\bar{\delta}}(\mathbf{r}, t) = \frac{1}{L^{3/2}}\sum_{\mathbf{k}}\hat{\bar{a}}_{\mathbf{k}}e^{i\mathbf{k}\cdot\mathbf{r}}, \tag{A.5}$$

$$\hat{\bar{a}}_{\mathbf{k}} = \frac{1}{L^{3/2}}\int d^3\mathbf{r}\,\hat{\bar{\delta}}(\mathbf{r}, t)e^{-i\mathbf{k}\cdot\mathbf{r}}, \tag{A.6}$$

where L is the side length of the finite quantization box in position space (with periodic boundary conditions) which is filled with the uniform source condensate, we may rewrite Eq. (A.4) as a pair of coupled differential equations in momentum space:

$$\frac{d\hat{\bar{a}}_{\mathbf{k}}}{dt} = -i\Delta_k\hat{\bar{a}}_{\mathbf{k}} - ig\hat{\bar{a}}_{-\mathbf{k}}^\dagger, \tag{A.7}$$

$$\frac{d\hat{\bar{a}}_{-\mathbf{k}}^\dagger}{dt} = i\Delta_k\hat{\bar{a}}_{-\mathbf{k}}^\dagger + ig\hat{\bar{a}}_{\mathbf{k}}. \tag{A.8}$$

Here we have defined the new variables $g = U\rho_0/\hbar$ and $\Delta_k = \hbar k^2/2m - \hbar k_0^2/2m$ for $k^2 \equiv |\mathbf{k}|^2$. These equations can be solved exactly and we transform back to the original frame to find

$$\hat{a}_\mathbf{k} = \alpha_\mathbf{k}(t)\hat{a}_\mathbf{k}(0) + \beta_\mathbf{k}(t)\hat{a}^\dagger_{-\mathbf{k}}(0), \tag{A.9}$$

$$\hat{a}^\dagger_{-\mathbf{k}} = \beta^*_\mathbf{k}(t)\hat{a}_\mathbf{k}(0) + \alpha^*_\mathbf{k}(t)\hat{a}^\dagger_{-\mathbf{k}}(0), \tag{A.10}$$

where $\hat{a}_\mathbf{k}$ is the Fourier component of the fluctuating field $\hat{\delta}(\mathbf{r}, t)$ and the time-dependent coefficients are

$$\alpha_\mathbf{k}(t) = \left[\cosh\left(\sqrt{g^2 - \Delta_k^2}\, t\right) + i\frac{\Delta_k}{\sqrt{g^2 - \Delta_k^2}}\sinh\left(\sqrt{g^2 - \Delta_k^2}\, t\right) \right] \exp\left(i\frac{\hbar k_0^2}{2m}t\right), \tag{A.11}$$

$$\beta_\mathbf{k}(t) = \frac{-ig}{\sqrt{g^2 - \Delta_k^2}}\sinh\left(\sqrt{g^2 - \Delta_k^2}\, t\right) \exp\left(i\frac{\hbar k_0^2}{2m}t\right). \tag{A.12}$$

The only non-zero expectation values (up to quadratic order) are then

$$n_\mathbf{k}(t) \equiv \langle \hat{a}^\dagger_\mathbf{k}(t)\hat{a}_\mathbf{k}(t)\rangle = |\beta_\mathbf{k}(t)|^2, \tag{A.13}$$

$$m_\mathbf{k}(t) \equiv \langle \hat{a}_\mathbf{k}(t)\hat{a}_{-\mathbf{k}}(t)\rangle = \alpha_\mathbf{k}(t)\beta_\mathbf{k}(t). \tag{A.14}$$

As the fluctuating component is initially vacuum and has Gaussian statistics we may invoke Wick's theorem to write any higher-order correlation function purely in terms of $n_\mathbf{k}(t)$ and $m_\mathbf{k}(t)$. Importantly, applying this to the back-to-back and collinear correlations gives the key results

$$g^{(2)}_{\mathrm{CL}} = g^{(2)}_{\mathbf{k},\mathbf{k}} = 2, \tag{A.15}$$

$$g^{(2)}_{\mathrm{BB}} = g^{(2)}_{\mathbf{k},-\mathbf{k}} = 2 + \frac{1}{n_\mathbf{k}}. \tag{A.16}$$

In the Gaussian approximation of Eqs. (1.38) and (1.39) this would correspond to $h_{\mathrm{CL}} = 1$ and $h_{\mathrm{BB}} = 1 + 1/n_\mathbf{k}$, respectively.

A.2 Perturbative Approach for Inhomogeneous Condensates in the Undepleted Pump Approximation

A limitation of the previous section was that the condensate was assumed to have a uniform density profile in position space, and thus reciprocally a delta-function distribution in momentum space. Such an assumption leads to only collisions between condensate atoms which have equal-but-opposite momenta. However, in realistic systems the initial condensate will have some finite momentum width, allowing collisions to occur between particles with momenta which are not directly opposing or of equal magnitude. Furthermore, unlike the uniform case, a collision between condensates with a inhomogeneous spatial profile will involve a decreasing overlap between the counter-propagating condensate wavepackets as they spatially separate. This should present itself as a time-dependent coupling between the condensate and scattered modes and will effect the dynamics and timescale of the scattering process. The method which we outline below allows us to capture these physical features. Importantly, the semi-analytic solutions which are derived are computationally far less intensive than the stochastic Bogoliubov approach employed extensively in Chaps. 2 and 3. In this respect they are an extremely useful tool with which to gain a qualitative understanding of spontaneous four-wave mixing in colliding condensates and in many cases (see, e.g., Ref. [3]) for quantitative predictions also.

As a first approximation, we assume that the initial condensate has a Gaussian density profile in position space such that the counter-propagating mean-field wavefunctions may be written as

$$\psi_{\pm k_0}(\mathbf{r}, t) = \sqrt{\frac{\rho_0}{2}} e^{-\frac{x^2}{2\sigma_x^2} - \frac{y^2}{2\sigma_y^2} - \frac{z^2}{2\sigma_y^2}} e^{\pm i k_0 \cdot \mathbf{r} - i \frac{\hbar |k_0|^2}{2m} t}. \tag{A.17}$$

The assumption of a Gaussian density profile is equivalent to assuming that the condensate is in the non-interacting ground state of a harmonic trapping potential. Realistically, for the trapping frequencies and interaction strength considered in this thesis the condensate density profile is better approximated by a Thomas-Fermi parabola. However, as will become clearer in the following sections, the Gaussian density profile allows somewhat simpler solutions of the scattering process whilst still encapsulating all the relevant physics.

In Chaps. 2 and 3 we discuss situations wherein the initial source condensate has an elongated (along the x direction) cigar-shaped density profile and is split radially into condensates with momenta $\pm k_0 \mathbf{e}_z$ respectively. For simplicity we will adopt this splitting geometry in the following solution. It is important to note that although the method we outline below is sufficiently general to apply to any splitting geometry, we make approximations which are only valid for our specific splitting scheme and thus affect the final form of the semi-analytic solutions.

Taking these approximations into account we may rewrite Eq. (A.4) in the form:

$$\frac{\partial \hat{\delta}(\mathbf{r}, t)}{\partial t} = -\frac{i\hbar}{2m} \nabla^2 \hat{\delta}(\mathbf{r}, t) + g(\mathbf{r}, t) \hat{\delta}^\dagger(\mathbf{r}, t), \tag{A.18}$$

where the time-varying and spatially dependent coupling to the source condensates is given by

$$g(\mathbf{r}, t) = 2U \psi_{+k_0}(\mathbf{r}, t) \psi_{-k_0}(\mathbf{r}, t),$$

$$= U\rho_0 e^{-iat} e^{-bt^2} e^{-\frac{x^2}{\sigma_x^2} - \frac{y^2}{\sigma_y^2} - \frac{z^2}{\sigma_y^2}}, \tag{A.19}$$

for $a = \hbar k_0^2/m$ and $b = \hbar^2 k_0^2/m^2\sigma_z^2$. From inspection of Eq. (A.19) we see there is a time-dependence of the overlap between the counter-propagating wavepackets which decays as a Gaussian whilst there is an overall phase-factor due to interference.

As previously, we can Fourier transform Eq. (A.18) and moving to a rotating frame $\hat{\bar{a}}(\mathbf{k}, t) = \hat{a}(\mathbf{k}, t)e^{i\hbar l^2 t/2m}$ where $\hat{a}(\mathbf{k}, t)$ is the momentum-space Fourier transform of $\hat{\delta}(\mathbf{r}, t)$ we have the operator equation of motion

$$\frac{d\hat{\bar{a}}(\mathbf{k}, t)}{dt} = \frac{-i}{\hbar} \int \frac{d\mathbf{q}}{(2\pi)^{3/2}} \tilde{g}(\mathbf{q} + \mathbf{k}, t)\hat{\bar{a}}^\dagger(\mathbf{q}, t)e^{i\frac{\hbar}{2m}(k^2+q^2)t}, \qquad (A.20)$$

where

$$\tilde{g}(\mathbf{k}, t) = \int \frac{d^3\mathbf{r}}{(2\pi)^{3/2}} e^{i\mathbf{k}\cdot\mathbf{r}} g(\mathbf{r}, t),$$

$$= \frac{U\rho_0}{2^{3/2}}\sigma_x\sigma_y\sigma_z e^{-iat-bt^2} e^{-\sum_i k_i^2\sigma_i^2/4}. \qquad (A.21)$$

is the Fourier transform of the coupling $g(\mathbf{r}, t)$. Substitution of this into Eq. (A.20) leads to the Heisenberg equation of motion

$$\frac{d\hat{\bar{a}}(\mathbf{k}, t)}{dt} = \mathcal{A}f(t) \int \frac{d\mathbf{q}}{(2\pi)^{3/2}} h(k, q, t)\hat{\bar{a}}^\dagger(\mathbf{q}, t), \qquad (A.22)$$

where $\mathcal{A} = -i(U\rho_0\sigma_x\sigma_y\sigma_z)/(\hbar 2^{3/2})$, $f(t) = e^{-iat-bt^2}$ and $h(k, q, t) = e^{\frac{i\hbar}{2m}(k^2+q^2)t-\sum_j(k_j+q_j)^2\sigma_j^2/4}$ for $j = x, y, z$. The form of this equation is illuminating when contrasted with that of Eqs. (A.7) and (A.8) for the uniform condensate. The finite momentum width of the source condensate in this approach leads to a momentum-dependent coupling to many momentum modes rather than purely between the $(\mathbf{k}, -\mathbf{k})$ pair.

In general, and particularly for the case of a Gaussian source condensate, this operator equation cannot be exactly solved. Instead, we adopt the approach outlined in Ref. [3] wherein the authors use a perturbative expansion in the momentum-space operator to derive approximate results for the expectation values $n(\mathbf{k}, \mathbf{k}') = \langle \hat{a}^\dagger(\mathbf{k}, t)\hat{a}(\mathbf{k}', t)\rangle$ and $m(\mathbf{k}, \mathbf{k}') = \langle \hat{a}(\mathbf{k}, t)\hat{a}(\mathbf{k}', t)\rangle$. The first step is to formally integrate Eq. (A.22),

$$\hat{\bar{a}}(\mathbf{k}, t) = \int_0^t d\tau \, \mathcal{A}f(\tau) \int \frac{d\mathbf{q}}{(2\pi)^{3/2}} h(k, q, \tau)\hat{\bar{a}}^\dagger(\mathbf{q}, \tau). \qquad (A.23)$$

The formal solution of this equation is expanded in a series of perturbative solutions $\hat{\bar{a}}(\mathbf{k}, t) = \sum_i \hat{\bar{a}}^{(i)}(\mathbf{k}, t)$ where to lowest order we have $\hat{\bar{a}}^{(0)}(\mathbf{k}, t) = \hat{\bar{a}}(\mathbf{k}, 0)$. Using this initial condition we can substitute $\hat{\bar{a}}(\mathbf{k}, t)$ back into Eq. (A.23) and solve iteratively to find, in the first order

$$\hat{\bar{a}}^{(1)}(\mathbf{k}, t) = \int_0^t d\tau \mathcal{A} f(\tau) \int \frac{d\mathbf{q}}{(2\pi)^{3/2}} h(k, q, \tau) \hat{\bar{a}}^{\dagger(0)}(\mathbf{q}, \tau). \tag{A.24}$$

A.2.1 Population of Collision Halo

Substituting the first-order perturbative result of Eq. (A.24) into the definition of $n(\mathbf{k}, \mathbf{k}')$ gives

$$n(\mathbf{k}, \mathbf{k}') = \langle \hat{\bar{a}}^{(1)\dagger}(\mathbf{k}, t) \hat{\bar{a}}^{(1)}(\mathbf{k}', t) \rangle e^{\frac{i\hbar}{2m}(k^2 - k'^2)t}. \tag{A.25}$$

The calculation of this quantity is straightforward yet lengthy. We present it in detail here as there are crucial differences to the derivation presented in Ref. [3]. These differences stem from our treatment of a collision along the radial axis of the condensates, whereas Ref. [3] addressed a collision along the longitudinal axis in comparison to earlier experimental results [7].

Using the definition of Eq. (A.24) we can rewrite Eq. (A.25) as

$$n(\mathbf{k}, \mathbf{k}') = e^{\frac{i\hbar}{2m}(k^2 - k'^2)t} \int_0^t d\tau \int_0^t d\tau' |\mathcal{A}|^2 f^*(\tau) f(\tau') \int \frac{d^3\mathbf{q}}{(2\pi)^3} h^*(k, q, \tau) h(k', q, \tau'),$$

$$= e^{\frac{i\hbar}{2m}(k^2 - k'^2)t} \int_0^t d\tau \int_0^t d\tau' |\mathcal{A}|^2 e^{ia(\tau - \tau') - b(\tau^2 + \tau'^2)}$$

$$\times \int \frac{d^3\mathbf{q}}{(2\pi)^3} e^{\frac{-i\hbar}{2m}(k^2 + q^2)\tau} e^{\frac{i\hbar}{2m}(k'^2 + q^2)\tau'} e^{-\sum_i (k_i + q_i)^2 \sigma_i^2 / 4} e^{-\sum_i (k_i' + q_i)^2 \sigma_i^2 / 4}. \tag{A.26}$$

The momentum space integral on the RHS of this equation,

$$I_{\mathbf{q}} = \int \frac{d^3\mathbf{q}}{(2\pi)^3} e^{\frac{-i\hbar}{2m}(k^2 + q^2)\tau} e^{\frac{i\hbar}{2m}(k'^2 + q^2)\tau'} e^{-\sum_i (k_i + q_i)^2 \sigma_i^2 / 4} e^{-\sum_i (k_i' + q_i)^2 \sigma_i^2 / 4}, \tag{A.27}$$

can be evaluated analytically. We present below a brief sketch of the solution and without loss of generality we present only the integral over q_x. Firstly we note we must complete the square in the Gaussian exponent,

$$\frac{-i\hbar}{2m}(k_x^2 + q_x^2)\tau + \frac{i\hbar}{2m}(k_x'^2 + q_x^2)\tau' - \frac{\sigma_x^2}{4}(k_x + q_x)^2 - \frac{\sigma_x^2}{4}(k_x' + q_x)^2$$

$$= -A_x(q_x - \frac{B_x}{2A})^2 + \frac{B_x^2}{4A_x} + \frac{i\hbar}{2m}(k_x'^2 \tau' - k_x^2 \tau) - \frac{\sigma_x^2}{4}(k_x + k_x'^2). \tag{A.28}$$

where $A_i = -i\hbar(\tau' - \tau)/2m + \sigma_i^2/2$ and $B_i = -\sigma_i^2(k_x + k_x')/2$ for $i = x, y, z$. We note further that

$$\frac{B_x^2}{4A_x} = \sigma_x^2/8(k_x + k_x')^2 + \frac{\frac{-i\hbar}{2m}(\tau - \tau')}{\frac{\sigma_x^2}{2} + \frac{i\hbar}{2m}(\tau - \tau')} \frac{\sigma_x^2}{8}(k_x + k_x')^2, \tag{A.29}$$

which implies Eq. (A.28) is equivalent to

$$-\frac{\sigma_x^2}{8}(k_x - k_x')^2 - \frac{\frac{i\hbar}{2m}(\tau - \tau')}{\frac{\sigma_x^2}{2} + \frac{i\hbar}{2m}(\tau - \tau')} \frac{\sigma_x^2}{8}(k_x + k_x')^2 + \frac{i\hbar}{2m}(k_x'^2\tau' - k_x^2\tau). \tag{A.30}$$

Thus the momentum-space integral over q_x becomes

$$I_{q_x} = \int \frac{dq_x}{2\pi} e^{\frac{i\hbar}{2m}(k_x'^2\tau' - k_x^2\tau)} e^{-A_x(q_x - \frac{B_x}{2A_x})^2} e^{-\frac{\sigma_x^2}{8}(k_x - k_x')^2} \exp\left[\frac{\frac{-i\hbar}{2m}(\tau - \tau')}{\frac{\sigma_x^2}{2} + \frac{i\hbar}{2m}(\tau - \tau')} \frac{\sigma_x^2}{8}(k_x + k_x')^2\right],$$

$$= \frac{e^{\frac{i\hbar}{2m}(k_x'^2\tau' - k_x^2\tau)}}{2\sqrt{\pi A_x}} e^{-\frac{\sigma_x^2}{8}(k_x - k_x')^2} \exp\left[\frac{\frac{-i\hbar}{2m}(\tau - \tau')}{\frac{\sigma_x^2}{2} + \frac{i\hbar}{2m}(\tau - \tau')} \frac{\sigma_x^2}{8}(k_x + k_x')^2\right]. \tag{A.31}$$

Performing identical integration over the remaining dimensions we can then substitute the result for $I_{\mathbf{q}}$ into Eq. (A.26) to give

$$n(\mathbf{k}, \mathbf{k}') = \frac{|\mathcal{A}|^2}{8\pi^{3/2}} e^{\frac{i\hbar}{2m}(k^2 - k'^2)t} e^{-\sum_i \frac{\sigma_i^2}{8}(k_i - k_i')^2} \int_0^t d\tau \int_0^t d\tau' e^{ia(\tau - \tau') - b(\tau^2 + \tau'^2)} e^{\frac{i\hbar}{2m}(k'^2\tau' - k^2\tau)}$$

$$\times \prod_i \left\{ \frac{1}{\sqrt{\frac{\sigma_i^2}{2} + \frac{i\hbar}{2m}(\tau - \tau')}} \exp\left[\frac{\frac{-i\hbar}{2m}(\tau - \tau')}{\frac{\sigma_i^2}{2} + \frac{i\hbar}{2m}(\tau - \tau')} \frac{\sigma_i^2}{8}(k_i + k_i')^2\right] \right\}. \tag{A.32}$$

To simplify the two-fold time integral we make a change of variables, $u = (\tau + \tau')/\sqrt{2}$ and $v = (\tau - \tau')/\sqrt{2}$, such that the first-order correlation can be written in terms of u and v as

$$n(\mathbf{k}, \mathbf{k}') = \frac{|\mathcal{A}|^2}{8\pi^{3/2}} e^{\frac{i\hbar}{2m}(k^2 - k'^2)t} e^{-\sum_i \frac{\sigma_i^2}{8}(k_i - k_i')^2}$$

$$\times \int_0^{\sqrt{2}t} du \int_{-u}^u dv e^{-b(u^2 + v^2)} e^{\frac{i\hbar}{2\sqrt{2}m}(k'^2 - k^2)u - \frac{i\hbar}{2\sqrt{2}m}(k'^2 + k^2)v + ia\sqrt{2}v}$$

$$\times \prod_i \left\{ \frac{1}{\sqrt{\frac{\sigma_i^2}{2} + \frac{i\hbar}{\sqrt{2}m}v}} \exp\left[\frac{\frac{-i\hbar}{\sqrt{2}m}v}{\frac{\sigma_i^2}{2} + \frac{i\hbar}{\sqrt{2}m}v} \frac{\sigma_i^2}{8}(k_i + k_i')^2\right] \right\}. \tag{A.33}$$

Next, by rationalizing the denominator of the exponents, Eq. (A.33) can be expressed as

$$n(\mathbf{k}, \mathbf{k}') = \frac{|\mathcal{A}|^2}{(2\pi)^{3/2}} e^{\frac{i\hbar}{2m}(k^2 - k'^2)t} \frac{e^{-\sum_i \frac{\sigma_i^2}{8}(k_i - k_i')^2}}{\sigma_x \sigma_y \sigma_z}$$

$$\times \int_0^{\sqrt{2}t} du \int_{-u}^{u} dv e^{-b(u^2 + vy^2)} e^{\frac{i\hbar}{2\sqrt{2}m}(k'^2 - k^2)u - \frac{i\hbar}{2\sqrt{2}m}(k'^2 + k^2)v + ia\sqrt{2}v}$$

$$\times \prod_i \left\{ \frac{1}{\sqrt{1 + \frac{i\hbar\sqrt{2}}{m\sigma_i^2}v}} \exp\left[\frac{-i\hbar}{\sqrt{2}m} v \frac{(k_i + k_i')^2}{4(1 + \frac{2\hbar^2}{m^2\sigma_i^4}v^2)} \right] \right. \tag{A.34}$$

$$\left. \times \exp\left[\frac{-\hbar^2}{m^2\sigma_i^2} v^2 \frac{(k_i + k_i')^2}{4(1 + \frac{2\hbar^2}{m^2\sigma_i^4}v^2)} \right] \right\}. \tag{A.35}$$

By considering the characteristic width of e^{-bv^2} we can assume that $\frac{2\hbar^2}{m^2\sigma_i^4}v^2 \ll 1$ and expand the exponents in a Taylor series to lowest order,

$$n(\mathbf{k}, \mathbf{k}') = \frac{|\mathcal{A}|^2}{(2\pi)^{3/2}} e^{\frac{i\hbar}{2m}(k^2 - k'^2)t} \frac{e^{-\sum_i \frac{\sigma_i^2}{8}(k_i - k_i')^2}}{\sigma_x \sigma_y \sigma_z}$$

$$\times \int_0^{\sqrt{2}t} du \int_{-u}^{u} dv e^{-b(u^2 + v^2)} e^{\frac{i\hbar}{2\sqrt{2}m}(k'^2 - k^2)u - \frac{i\hbar}{2\sqrt{2}m}(k'^2 + k^2)v + ia\sqrt{2}v}$$

$$\times \prod_i \frac{1}{\sqrt{1 + \frac{i\hbar\sqrt{2}}{m\sigma_i^2}v}} \exp\left[\frac{-i\hbar}{\sqrt{2}m} v \frac{(k_i + k_i')^2}{4}\left(1 - \frac{2\hbar^2}{m^2\sigma_i^4}v^2\right) \right]$$

$$\times \exp\left[\frac{-\hbar^2}{m^2\sigma_i^2} v^2 \frac{(k_i + k_i')^2}{4}\left(1 - \frac{2\hbar^2}{m^2\sigma_i^4}v^2\right) \right]. \tag{A.36}$$

Ignoring higher-order terms in the expansion the integral may be approximated as

$$n(\mathbf{k}, \mathbf{k}') = \frac{|\mathcal{A}|^2}{(2\pi)^{3/2}} e^{\frac{i\hbar}{2m}(k^2 - k'^2)t} \frac{e^{-\sum_i \frac{\sigma_i^2}{8}(k_i - k_i')^2}}{\sigma_x \sigma_y \sigma_z} \int_0^{\sqrt{2}t} du \int_{-u}^{u} dv e^{-bu^2} e^{\frac{i\hbar}{2\sqrt{2}m}(k'^2 - k^2)u}$$

$$\times \exp\left[\frac{-i\hbar}{2m} v \left(-2\sqrt{2}k_0^2 + \frac{k^2 + k'^2}{\sqrt{2}} + \sum_i \frac{(k_i + k_i')^2}{2\sqrt{2}} \right) \right]$$

$$\times \exp\left[\frac{-\hbar^2}{m^2} v^2 \left(\frac{k_0^2}{\sigma_z^2} + \sum_i \frac{(k_i + k_i')^2}{4\sigma_i^2} \right) \right],$$

$$= \frac{|\mathcal{A}|^2}{(2\pi)^{3/2}} \frac{e^{-\sum_i \frac{\sigma_i^2}{8}(k_i - k_i')^2}}{\sigma_x \sigma_y \sigma_z} \int_0^{\sqrt{2}t} du \, e^{-bu^2} e^{\frac{i\hbar}{2\sqrt{2}m}(k'^2 - k^2)u}$$

$$\times \sqrt{\frac{\pi}{\delta} \frac{m}{\hbar}} e^{-\gamma^2/16\delta} \mathrm{Re}\left[\mathrm{erf}\left(\frac{\hbar}{m}\sqrt{\delta}u + i\frac{\gamma}{4\sqrt{\delta}} \right) \right], \tag{A.37}$$

where

$$\gamma = -2\sqrt{2}k_0^2 + \frac{k^2 + k'^2}{\sqrt{2}} + \sum_i \frac{(k_i + k_i')^2}{2\sqrt{2}}, \tag{A.38}$$

$$\delta = \frac{k_0^2}{\sigma_z^2} + \sum_i \frac{(k_i + k_i')^2}{4\sigma_i^2}. \tag{A.39}$$

The final integration over u must be performed numerically as it is analytically intractable. This is different to the result arrived at in Ref. [3], as the orientation of their collision led to a significantly different characteristic width of the function in the u and v variables. This allowed the final pair of integrals to decouple and be performed exactly.

Lastly, to calculate the momentum-space population density in the collision halo we set $\mathbf{k}' = \mathbf{k}$ in $n(\mathbf{k}, \mathbf{k}')$ to give

$$n(\mathbf{k}) = \frac{|\mathcal{A}|^2 e^{-\gamma^2/16\delta}}{(2\pi)^{3/2}\sigma_x\sigma_y\sigma_z}\sqrt{\frac{\pi}{\delta}}\frac{m}{\hbar}\int_0^{\sqrt{2}t} du \, e^{-bu^2} \mathrm{Re}\left[\mathrm{erf}\left(\frac{\hbar}{m}\sqrt{\delta}u + i\frac{\gamma}{4\sqrt{\delta}}\right)\right]. \tag{A.40}$$

A.2.2 Anomalous Moment

Similarly, substitution of the perturbative series into the definition of the anomalous moment $m(\mathbf{k}, \mathbf{k}')$ gives to lowest order

$$m(\mathbf{k}, \mathbf{k}') = e^{\frac{i\hbar}{2m}(k^2 + k'^2)t}\langle \hat{a}^{(0)}(\mathbf{k}, t)\hat{a}^{(1)}(\mathbf{k}', t)\rangle, \tag{A.41}$$

$$= A\int_0^t d\tau f(\tau)\int \frac{d\mathbf{q}}{(2\pi)^{3/2}}h(k', q, \tau)\langle \hat{a}(\mathbf{k}, 0)\hat{a}^\dagger(\mathbf{q}, 0)\rangle. \tag{A.42}$$

Using the bosonic commutation relations $[\hat{a}(\mathbf{k}, 0), \hat{a}^\dagger(\mathbf{k}', 0)] = \delta(\mathbf{k} - \mathbf{k}')$ we can simplify this to (for a vacuum initial condition, as is the case assumed here):

$$m(\mathbf{k}, \mathbf{k}') = e^{\frac{i\hbar}{2m}(k^2 + k'^2)t}\frac{A}{(2\pi)^{3/2}}\int_0^t d\tau e^{-ia\tau - b\tau^2}e^{\frac{i\hbar}{2m}(k^2+k'^2)\tau}e^{-\sum_i(k_i+k_i')^2\sigma_i^2/4}. \tag{A.43}$$

Completing the square in τ and using a change of variables allows one to evaluate the integral explicitly to arrive at the final result

$$m(\mathbf{k}, \mathbf{k}') = \frac{A}{4\pi\sqrt{b}}e^{-\Delta^2/4b}e^{-\sum_i(k_i+k_i')^2\sigma_i^2/4}e^{\frac{i\hbar}{2m}(k^2+k'^2)t}\left[\mathrm{erf}\left(\frac{i\Delta}{2\sqrt{b}}\right) + \mathrm{erf}\left(\sqrt{b}t - \frac{i\Delta}{2\sqrt{b}}\right)\right], \tag{A.44}$$

where

$$\Delta = \frac{\hbar}{m} \left[\frac{k^2 + k'^2}{2} - k_0^2 \right].$$ (A.45)

To benchmark the results of the perturbative method we compare the results to both the homogeneous approach of the previous section and a full numerical simulation utilizing the stochastic Bogoliubov method (for further details see Chaps. 2 and 3). We note that the results for the stochastic Bogoliubov method are based upon a different initial source condensate, whose wavefunction is found by an imaginary-time numerical calculation. However, this difference does not change the qualitative features which we compare between the models. To ensure the results are quantitatively similar we fix the peak density of the imaginary-time and Gaussian solutions to be identical and then fix the rms widths of the Gaussian by a best-fit to the density profile of the imaginary-time result. Similarly, the uniform condensate for the homogeneous approach is constructed by fixing the total number of particles in the quantization box to match the imaginary time solution, whilst the box size L is approximately the same size as the trapped condensate. Explicitly we choose $L = 2R_{TF}$ where R_{TF} is the Thomas-Fermi radius of the condensate and is a good approximation to the imaginary-time calculation.

In Fig. A.1 we compare various properties of the scattering halo population $n(\mathbf{k}) \equiv n(\mathbf{k}, \mathbf{k})$. In particular, we characterize the radial density profile which is approximated well by the Gaussian function $n(k_r, \phi) = n_p(\phi)\exp(-(k_r - k_p)^2/2(\delta k_r)^2)$ where n_p is the peak density, k_p is the peak radius and δk_r is the root-mean-square (rms) width of the halo. The analytic and numerical methods employing inhomogeneous condensates show good qualitative agreement in terms of the asymmetry of collision halo, in particular the oscillations in halo width and peak density. These features are partly a consequence of the anisotropy of the initial atomic cloud and are discussed in further detail in Ref. [8]

An important physical difference to the perturbative and homogeneous techniques is that the stochastic Bogoliubov method incorporates the mean-field potential felt by the scattered atoms due to the source condensates, an effect which is not included in Eqs. (A.4) or (A.18). A consequence of excluding the mean-field potential of the condensates is most starkly evidenced in Fig. A.1a, c, wherein we note the discrepancy in the peak-radius of the collision halo. This result is intuitive for the perturbative and homogeneous treatments as the requirement of energy conservation translates to the kinetic energy of the colliding atomic pair $E_i \simeq \hbar k_0^2/m$ being equal to that of the final pair, such that we expect the peak to occur at $k_r = k_0$. In contrast, in the stochastic Bogoliubov treatment the initial energy of the colliding pair is made up of kinetic *and* mean-field components. The effective mean-field potential felt by the scattered atoms is thus increased and consequentially the scattered atoms will have a smaller outgoing momentum than that of the ingoing pair, implying $k_r < k_0$ [2].

Although not a feature in the collision geometry we discuss here and in Chaps. 2 and 3 (split along the tight trapping direction, corresponding to the z axis here), it is also worth noting the importance of the more complicated time-evolution of the source condensates. According to mean-field GPE evolution, the elongated source

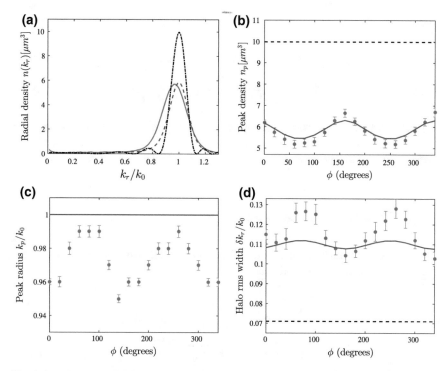

Fig. A.1 **a** Average radial density profile of collision halo from stochastic Bogoliubov simulations (*red solid line*), perturbative solution [Eq. (A.25), *blue solid line*] and homogeneous condensate solution (*black solid line*) in the k_x–k_y cross-section. **b** Comparison of oscillation in peak density of the collision halo n_p as a function of angle ϕ. **c** Variation of peak radius of the collision halo k_p as a function of ϕ. **d** RMS width of the collision halo δk_r as a function of ϕ. For **b**–**d** we compare the stochastic Bogoliubov calculation (*red circles*), perturbative solution (*blue solid line*) and homogeneous condensate solution (*black solid line*) for equivalent source condensates (see text) and collision duration (color figure online)

condensates will expand asymmetrically after the trapping potential is removed. In earlier experimental schemes [1, 9] the collision took place along the elongated x-axis of the condensate, involving much longer timescales before the counter-propagating condensates spatially separated [1]. Indeed, for this geometry the relevant timescale of the collision process is given by the reduction in peak density of the condensates $\rho(t) = |\psi(\mathbf{r}, t)|^2$ as the clouds expand rapidly along the radial axes. Clearly this feature would not be replicated by the simple approximation of $\psi(\mathbf{r}, t)$ given by Eq. (A.17). However, it is possible to further adapt the perturbative technique by using the scaling solutions for the self-similar expansion of a condensate outlined in Ref. [10]. This leads to an improved form of the split wavepackets

$$\psi_{\pm \mathbf{k}_0}(\mathbf{r}, t) = \sqrt{\frac{\rho_0(t)}{2}} e^{-\frac{x^2}{2\sigma_x^2} - \frac{y^2}{2\sigma_y^2(t)} - \frac{z^2}{2\sigma_z^2(t)}} e^{\pm i \mathbf{k}_0 \cdot \mathbf{r} - i \frac{\hbar |\mathbf{k}_0|^2}{2m} t}. \tag{A.46}$$

where

$$\sigma_{y,z}(t) = \sqrt{1 + (\omega_{y,z}t)^2}\sigma_{y,z}(0), \tag{A.47}$$

$$\rho_0(t) = \frac{\rho_0(0)}{\sqrt{(1 + (\omega_y t)^2)(1 + (\omega_z t)^2)}}, \tag{A.48}$$

and we assume the expansion along the x-axis is negligible compared to the y and z directions. Solving the ensuing operator equations follows the same approach as outlined above, however, we do not present any results for this in this appendix.

A.2.3   Second-Order Correlations

The back-to-back and collinear correlations can be calculated from the first-order moments by invoking Wick's theorem to give:

$$g_{CL}^{(2)}(\mathbf{k}, \mathbf{k}') = 1 + \frac{\left[n(\mathbf{k}, \mathbf{k}')\right]^2}{n(\mathbf{k})n(\mathbf{k}')}, \tag{A.49}$$

$$g_{BB}^{(2)}(\mathbf{k}, \mathbf{k}') = 1 + \frac{|m(\mathbf{k}, \mathbf{k}')|^2}{n(\mathbf{k})n(\mathbf{k}')}. \tag{A.50}$$

where $\mathbf{k}' \simeq \mathbf{k}$ and $\mathbf{k}' \simeq -\mathbf{k}$, respectively. Comparing to the Gaussian approximation of Eqs. (1.38) and (1.39) we trivially recognize $h_{CL} = 1$ and $h_{BB} = |m(\mathbf{k}, -\mathbf{k})|^2/[n(\mathbf{k})n(-\mathbf{k})]$.

Although the analytic solution of $g_{CL}^{(2)}(\mathbf{k}, \mathbf{k}')$ requires the remaining integral of Eq. (A.37) to be calculated using numerical techniques, it is still possible to extract a good estimate of the correlation width by inspection of the remainder of Eq. (A.37). We ignore the terms within the integrand as they depend only on the sums $\mathbf{k} + \mathbf{k}'$ and $k^2 + k'^2$ respectively, whereas the correlation width will most sensitively depend on the difference $\Delta\mathbf{k} = \mathbf{k} - \mathbf{k}'$ for $\mathbf{k}' \simeq \mathbf{k}$. It is then obvious to identify the Gaussian dependence $n(\mathbf{k}, \mathbf{k} + \Delta\mathbf{k}) \propto \exp[-\sum_i(\Delta k_i)^2\sigma_i^2/8]$ and thus we can show that this leads to $\sigma_{CL} = \sqrt{2}/\sigma_i = 2\sqrt{2}\sigma_{k_i}$ where σ_{k_i} for $k_i = k_x, k_y, k_z$ is the rms width of the momentum-space wavefunction of the split condensates. This correlation width is found to be consistent with the analytic solution found when Eq. (A.37) is numerically evaluated.

Similarly, for the back-to-back correlation we may identify the momentum-dependent Gaussian decay in Eq. (A.44) as $m(\mathbf{k}, -\mathbf{k}+\Delta\mathbf{k}) \propto \exp[-\sum_i(\Delta k_i)^2\sigma_i^2/4]$ and thus we associate a related back-to-back correlation width of $\sigma_{BB,i} = 1/\sigma_i = 2\sigma_{k_i}$. Comparing these correlations widths we find a ratio of $\sigma_{CL}/\sigma_{BB} = \sqrt{2}$, which is consistent with previous calculations where the initial condensate is approximated as a Gaussian [5].

A.3 Conclusion

The previous sections demonstrate how many of the phenomena present in the spontaneous four-wave mixing process can be encapsulated by simple models. The simplest homogeneous model is used extensively throughout Chaps. 2 and 3 as a basic 'toy model' of the process. In particular, in the supplementary material of Chap. 2 (found in Appendix C) the model is able to predict many features of the Hong–Ou–Mandel effect such as the profile of the infamous Hong–Ou–Mandel 'dip'. Similarly, the perturbative approach is an essential tool for more in-depth analysis of the collision process, without having to resort to the computationally intensive stochastic Bogoliubov approach. Specifically, the analytic form of the anomalous moment allowed an accurate treatment of the effects of phase dispersion between scattered pairs in the collision halo, which is discussed in more detail in the supplementary material of Chap. 2 (found in Appendix D).

References

1. Perrin, A., et al.: Observation of atom pairs in spontaneous four-wave mixing of two colliding Bose-Einstein condensates. Phys. Rev. Lett. **99**, 150405 (2007)
2. Krachmalnicoff, V., et al.: Spontaneous four-wave mixing of de Broglie waves: beyond optics. Phys. Rev. Lett. **104**, 150402 (2010)
3. Chwedeńczuk, J., et al.: Pair correlations of scattered atoms from two colliding Bose-Einstein condensates: Perturbative approach. Phys. Rev. A **78**, 053605 (2008)
4. Deuar, P., Chwedeńczuk, J., Trippenbach, M., Ziń, P.: Bogoliubov dynamics of condensate collisions using the positive-P representation. Phys. Rev. A **83**, 063625 (2011)
5. Ogren, M., Kheruntsyan, K.V.: Atom–atom correlations in colliding Bose-Einstein condensates. Phys. Rev. A. **79**, 021606 (2009)
6. Ziń, P., Chwedeńczuk, J., Trippenbach, M.: Elastic scattering losses from colliding Bose-Einstein condensates. Phys. Rev. A **73**, 033602 (2006)
7. Perrin, A., et al.: Atomic four-wave mixing via condensate collisions. New J. Phys. **10**, 045021 (2008)
8. Deuar, P., et al.: Anisotropy in s-wave Bose-Einstein condensate collisions and its relationship to superradiance. Phys. Rev. A **90**, 033613 (2014)
9. Perrin, A., et al.: Atomic four-wave mixing via condensate collisions. New J. Phys. **10**, 045021 (2008)
10. Castin, Y., Dum, R.: Bose-Einstein condensates in time dependenttraps. Phys. Rev. Lett. **77**, 5315–5319 (1996)

Appendix B
Mean-Field Theory of Bragg Scattering

B.1 Basic Theory and Understanding

In atom optics, the role of optical mirrors and beam-splitters is taken by Bragg pulses. They play a crucial role in allowing us to implement atomic analogs of the Rarity–Tapster and Hong–Ou–Mandel interferometers. In their simplest form, the behaviour of such pulses is well known [1], however, there are several key issues which we wish to build a theoretical understanding of. Firstly, we must understand how the choice of experimental parameters, such as laser intensity, affect the choice of scattering regime; and secondly, we seek to investigate the effect of Bragg pulses implemented on realistic systems of particles which have a finite spread of momentum.

Bragg pulses are created by counter-propagating laser beams which form a periodic optical potential. Atoms passing through this potential can be scattered, similar to light passing through a diffraction grating in space. Fundamentally, the scattering process allows the manipulation and transfer of the atom(s) between different momentum states. As will become clear in the following, careful control of the intensity and applied duration of the laser beams allows one to control the portion of atoms transferred between momentum states. Formally, this diffraction of matter-waves is known as Kapitza-Dirac scattering [2] and can result in multiple scattering resonances. However, for our purposes we will consider only the special case of Bragg scattering [1], where only the initial momentum mode and another single well-defined momentum mode are involved.

An illustration of the process is given in Fig. B.1. The periodic potential is formed by two counter-propagating laser beams with wave-vectors $\mathbf{k}_{L,1}$ and $\mathbf{k}_{L,2}$ respectively. In the simplest case of a standing-wave (stationary) optical lattice we have $\mathbf{k}_{L,1} = -\mathbf{k}_{L,2} \equiv \mathbf{k}_L$ where $\mathbf{k}_L$ is the wavevector of the lattice. If we consider an atom (in its electronic ground state) with initial momentum $\mathbf{k}_1 (\equiv -\mathbf{k}_{L,1})$ and energy E_g subject to this optical potential, it may absorb a photon from the laser with wavevector $\mathbf{k}_{L,1}$ and move to a short-lived excited state (with atomic momentum $\mathbf{k} = 0$) detuned by Δ from the excited state with energy E_e. Due to the presence of the second laser

© Springer International Publishing Switzerland 2016
R.J. Lewis-Swan, *Ultracold Atoms for Foundational Tests of Quantum Mechanics*, Springer Theses, DOI 10.1007/978-3-319-41048-7

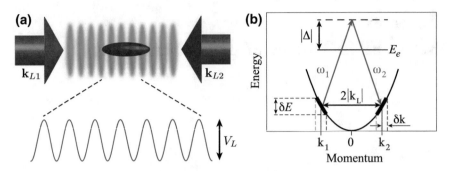

Fig. B.1 **a** Optical lattice created by two counter-propagating laser beams with wavevectors $\mathbf{k}_{L,1} \equiv -\mathbf{k}_1$ and $\mathbf{k}_{L,2} \equiv -\mathbf{k}_2$. Adopted from Ref. [3]. **b** Schematic diagram of Bragg scattering process for a stationary Bragg pulse ($\mathbf{k}_1 = -\mathbf{k}_2$). An atom (initially in its electronic ground state) with momentum $\mathbf{k}_1$ absorbs a photon with energy $\hbar\omega_1$ and moves to an excited state detuned by $\hbar\Delta$ from the energy level E_e. The atom then undergoes stimulated emission and emits a photon of energy $\hbar\omega_2$. The net momentum change of $2\mathbf{k}_L$ leaves the atom in the $\mathbf{k}_2$ momentum mode. The quadratic curve indicates the atomic dispersion relation $E = \hbar^2 |\mathbf{k}|^2 / 2m$. Off-resonant coupling is indicated by $\delta\mathbf{k}$ and consequential energy mismatch δE

beam, the atom can undergo stimulated emission and emit a photon with wavevector $\mathbf{k}_{L,2}$. The atom thus undergoes a change in momentum of $\mathbf{k}_{L,2} - \mathbf{k}_{L,1} = 2\mathbf{k}_L$, and is transferred to a state with final momentum $\mathbf{k}_2 (\equiv -\mathbf{k}_{L,2})$ For a stationary lattice (standing wave potential) it is important to note that the energy is unchanged between the initial and final state,

$$\delta E = \frac{\hbar|\mathbf{k}_1|^2}{2m} - \frac{\hbar|\mathbf{k}_2|^2}{2m} = 0. \tag{B.1}$$

In the following we drop the notational distinction between the photon wavevector and atomic momenta and simply refer to $\mathbf{k}_1$ and $\mathbf{k}_2$ which are equivalent for both.

An important question is what occurs when an atom with $\mathbf{k}_1' = \mathbf{k}_1 + \delta\mathbf{k}$ enters the potential. This is particularly important as one in general deals with an ensemble of atoms such as a BEC with some finite momentum width. By momentum conservation one predicts the mode $\mathbf{k}_1'$ may couple to $\mathbf{k}_2' = \mathbf{k}_2 + \delta\mathbf{k}$, however, this will not conserve energy,

$$\delta E = \frac{\hbar|\mathbf{k}_1|^2}{2m} - \frac{\hbar|\mathbf{k}_2|^2}{2m} \neq 0. \tag{B.2}$$

A simple heuristic argument can be constructed to demonstrate that such off-resonant coupling can occur under certain conditions. This argument, based on energy-time uncertainty, was originally presented in Ref. [2]. The uncertainty in the energy of the interaction between the atom and the optical lattice is defined as ΔE, and the applied duration of the potential Δt is taken to be that required for a π- or $\pi/2$-pulse. The energy-time uncertainty gives the relation $\Delta E \Delta t \geq \hbar$ and thus $\Delta E \geq \hbar/\delta t$. If the energy difference between the coupled momentum states δE is of the same magnitude as the energy uncertainty ΔE then the modes $\mathbf{k}_1'$ and $\mathbf{k}_2'$ will couple. Effectively, this means that short pulses lead to a broad range of momentum states being coupled, whereas long pulses are selective and only couple narrow momentum ranges.

A similar argument is used to ensure the validity of assuming the Bragg scattering regime. By ensuring the pulse duration is sufficiently long that $\Delta E \simeq \hbar/\delta t \ll \hbar|\mathbf{k}_1|^2/2m$ the atoms are unlikely to undergo multiple absorption/emission events to higher momentum modes due to conservation of energy. A secondary requirement is that the strength of the optical lattice is weak compared to the recoil energy, $V_L = |\Omega|^2/2\Delta \ll \hbar|\mathbf{k}_1|^2/2m$ where Ω is the intensity of the laser beams, such that the atoms only weakly interact with the potential [2, 3].

B.2 Analytic Solution of Bragg Pulse Transformation

The behaviour of a BEC in an optical lattice potential is given by the effective non-interacting Hamiltonian [3]

$$\hat{H} = \int d^3 \mathbf{r} \hat{\psi}^\dagger(\mathbf{r}) \left[\left(\frac{-\hbar^2}{2m} \nabla^2 \right) + \frac{i|\Omega|^2}{2\Delta} \cos\left([\mathbf{k}_2 - \mathbf{k}_1] \cdot \mathbf{r} + \delta t + \theta \right) \right] \hat{\psi}(\mathbf{r}), \quad \text{(B.3)}$$

where Ω is the Rabi frequency (determined by the laser intensity), $\mathbf{k}_{1,2}$ the respective wavevectors of the laser beams, $\delta = \omega_1 - \omega_2$ the difference of their frequencies, θ their relative phase and $\Delta = E_e/\hbar - \omega_1$ is the detuning of the laser induced transition from the excited state and we arbitrarily set $E_g = 0$.

Considering a stationary optical lattice, for which $\delta = 0$, the Heisenberg operator equation for $\hat{\psi}(\mathbf{r})$ is then,

$$\frac{d\hat{\psi}(\mathbf{r}, t)}{dt} = \frac{i\hbar}{2m} \nabla^2 \hat{\psi}(\mathbf{r}, t)$$
$$+ \frac{i|\Omega|^2}{2\Delta} \cos\left([\mathbf{k}_2 - \mathbf{k}_1] \cdot \mathbf{r} + \theta \right) \hat{\psi}(\mathbf{r}, t). \quad \text{(B.4)}$$

To solve this equation we first make the mean-field approximation $\hat{\psi}(\mathbf{r}) \to \psi(\mathbf{r})$ and then take the Fourier transform of Eq. (B.4),

$$\frac{d\psi(\mathbf{k}, t)}{dt} = \frac{-i\hbar|\mathbf{k}|^2}{2m} \psi(\mathbf{k}, t) - \frac{i\Omega_{\text{eff}}}{2} \left[ie^{i\theta} \psi(\mathbf{k} + 2\mathbf{k}_L, t) \right.$$
$$\left. - ie^{-i\theta} \psi(\mathbf{k} - 2\mathbf{k}_L, t) \right], \quad \text{(B.5)}$$

where $2\mathbf{k}_L = \mathbf{k}_2 - \mathbf{k}_1$ and $\Omega_{\text{eff}} = |\Omega|^2/2\Delta$ is the effective Rabi frequency. As we are considering scattering in the Bragg regime, we consider coupling only around the two momentum states $\mathbf{k}_1 = -\mathbf{k}_2 = \mathbf{k}_L$. Under this condition, Eq. (B.5) can be rewritten in the form of two coupled equations,

$$\frac{d\psi(\mathbf{k}'_1, t)}{dt} = \frac{-i\hbar|\mathbf{k}'_1|^2}{2m}\psi(\mathbf{k}'_1, t) + i\frac{\Omega_{\text{eff}}}{2}\psi(\mathbf{k}'_2, t) \tag{B.6}$$

$$\frac{d\psi(\mathbf{k}'_2, t)}{dt} = \frac{-i\hbar|\mathbf{k}'_2|^2}{2m}\psi(\mathbf{k}'_2, t) + i\frac{\Omega_{\text{eff}}}{2}\psi(\mathbf{k}'_1, t) \tag{B.7}$$

where $\mathbf{k}'_1 = \mathbf{k}_1 + \delta\mathbf{k}$ and $\mathbf{k}'_2 = \mathbf{k}_2 + \delta\mathbf{k}$ as previously defined and we assume $\theta = \pi/2$ for definiteness.

These coupled differential equations can readily be solved to give

$$\psi(\mathbf{k}'_1, t) = A(\mathbf{k}_L, \delta k, t)\psi(\mathbf{k}'_1, 0) + B(\mathbf{k}_L, \delta k, t)\psi(\mathbf{k}'_2, 0), \tag{B.8}$$

$$\psi(\mathbf{k}'_2, t) = C(\mathbf{k}_L, \delta k, t)\psi(\mathbf{k}'_1, 0) + D(\mathbf{k}_L, \delta k, t)\psi(\mathbf{k}'_2, 0), \tag{B.9}$$

where,

$$A(\mathbf{k}_L, \delta k, t) = \exp\left[\frac{-i\hbar}{2m}\left(|\mathbf{k}_L|^2 + \delta k^2\right)t\right]$$
$$\times \left[\frac{i\alpha}{\sqrt{\alpha^2 + \Omega_{\text{eff}}^2}}\sin\left(\frac{1}{2}\sqrt{\alpha^2 + \Omega_{\text{eff}}^2}t\right) + \cos\left(\frac{1}{2}\sqrt{\alpha^2 + \Omega_{\text{eff}}^2}t\right)\right], \tag{B.10}$$

$$B(\mathbf{k}_L, \delta k, t) = \frac{i\Omega_{\text{eff}}}{\sqrt{\alpha^2 + \Omega_{\text{eff}}^2}}\exp\left[\frac{-i\hbar}{2m}\left(|\mathbf{k}_L|^2 + \delta k^2\right)t\right]\sin\left(\frac{1}{2}\sqrt{\alpha^2 + \Omega_{\text{eff}}^2}t\right), \tag{B.11}$$

$$C(\mathbf{k}_L, \delta k, t) = \frac{i\Omega_{\text{eff}}}{\sqrt{\alpha^2 + \Omega_{\text{eff}}^2}}\exp\left[\frac{-i\hbar}{2m}\left(|\mathbf{k}_L|^2 + \delta k^2\right)t\right]\sin\left(\frac{1}{2}\sqrt{\alpha^2 + \Omega_{\text{eff}}^2}t\right), \tag{B.12}$$

$$D(\mathbf{k}_L, \delta k, t) = \exp\left[\frac{-i\hbar}{2m}\left(|\mathbf{k}_L|^2 + \delta k^2\right)t\right]$$
$$\times \left[\frac{-i\alpha}{\sqrt{\alpha^2 + \Omega_{\text{eff}}^2}}\sin\left(\frac{1}{2}\sqrt{\alpha^2 + \Omega_{\text{eff}}^2}t\right) + \cos\left(\frac{1}{2}\sqrt{\alpha^2 + \Omega_{\text{eff}}^2}t\right)\right], \tag{B.13}$$

where $\alpha = 4\hbar^2(\mathbf{k}_L \cdot \delta\mathbf{k})^2/m^2$. In the resonant case, $\delta\mathbf{k} = 0$, the solutions take the simple form

$$\psi(\mathbf{k}_1, t) = \exp\left[\frac{-i\hbar|\mathbf{k}_L|^2}{2m}t\right]\left[\cos\left(\frac{\Omega_{\text{eff}}t}{2}\right)\psi(\mathbf{k}_1, 0) + \sin\left(\frac{\Omega_{\text{eff}}t}{2}\right)\psi(\mathbf{k}_2, 0)\right],$$

(B.14)

$$\psi(\mathbf{k}_2, t) = \exp\left[\frac{-i\hbar|\mathbf{k}_L|^2}{2m}t\right]\left[\sin\left(\frac{\Omega_{\text{eff}}t}{2}\right)\psi(\mathbf{k}_1, 0) + \cos\left(\frac{\Omega_{\text{eff}}t}{2}\right)\psi(\mathbf{k}_2, 0)\right].$$

(B.15)

By examination of Eqs. (B.14) and (B.15) we can readily associate a beam-splitter transformation (or $\pi/2$-pulse) with $\Omega_{\text{eff}}t_{\pi/2} = \pi/2$ and a mirror transformation (or π-pulse) with $\Omega_{\text{eff}}t_\pi = \pi$.

Examining the full solutions [Eqs. (B.10)–(B.13)] we see that coupling between off-resonant modes is equivalent to an increase of the effective Rabi frequency $\Omega'_{\text{eff}} \equiv \sqrt{\alpha^2 + \Omega_{\text{eff}}^2}$. Hence, whilst the 'targeted' resonant modes may transform according to the canonical definition of the π- and $\pi/2$-pulses, a finite shift δk (along the direction of the Bragg vector $\mathbf{k}_L$) will lead to imperfect transfer (i.e., not 100–0 or 50–50 respectively) between the coupled modes. Naively, one would assume that the solution to this issue would be to satisfy the condition $\Omega_{\text{eff}}^2 \gg \alpha^2$ for all relevant momenta. However, increasing Ω_{eff} would also lead to increased coupling to higher momentum modes, i.e., increasing the chance of multiple pairs of scattering events, which would push the scattering out of the Bragg regime and hence also degrade the efficiency of the π- and $\pi/2$-pulses.

Effects due to off-resonant coupling play a key role in Chaps. 2 and 3 and have previously been qualitatively studied in Ref. [4] in the context of atom interferometry.

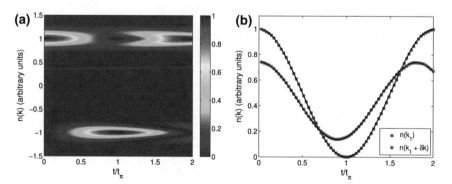

Fig. B.2 **a** Momentum distribution $n(k)$ of a 1D BEC as a function of Bragg pulse duration t. The inhomogeneous transfer of population between the coupled momentum modes is clearly illustrated. **b** Population of resonant [$n(k_1)$, *blue squares*] and off-resonant [$n(k_1 + \delta k)$ for $\delta k = 0.08|k_L|$, *red circles*] momentum states as a function of pulse duration. Excellent agreement is found between the analytic model of Eqs. (B.8) and (B.9) [*solid black lines*], which only assumes the coupling is limited to between two modes, and numerical solution of Eq. (B.4) in the mean-field approximation (*blue* and *red squares*), which places no limit on the number of modes which couple [including kicks to $k_1 + 2|k_0|$ and $k_2 - 2|k_0|$, which are included in the simulation lattice, though not shown in (**a**)]

We illustrate the effect of coupling between off-resonant modes in Fig. B.2, where we numerically simulate an example Bragg pulse [i.e., we take the mean-field approximation of Eq. (B.4) with no further approximations] for a 1D BEC. It is clear that while the central portion of the atomic cloud is coupled to the new momentum mode, the edges of the distribution are not transferred efficiently as they are off-resonant.

References

1. Meystre, P.: Atom Optics. Springer, New York (2001)
2. Batelaan, H.: The Kapitza-Dirac effect. Contemp. Phys. **41**, 369–381 (2000)
3. Ferris, A.J.: Thermalisation, correlations and entanglement in Bose-Einstein condensates. Ph.D. thesis, University of Queensland (2009)
4. Szigeti, S.S., Debs, J.E., Hope, J.J., Robins, N.P., Close, J.D.: Why momentum width matters for atom interferometry with Bragg pulses.New J. Phys. **14**, 023009 (2012)

Appendix C
Supplementary Material for Chapter 2

C.1 Methods

To simulate the collision dynamics, we use the time-dependent stochastic Bogoliubov approach [1, 2] used previously to accurately model a number of condensate collision experiments [1, 3, 4]. In this approach, the atomic field operator is split into $\hat{\psi}(\mathbf{r}, t) = \psi(\mathbf{r}, t) + \hat{\delta}(\mathbf{r}, t)$, where ψ is the mean-field component describing the source condensates and $\hat{\delta}$ is the fluctuating component (treated to lowest order in perturbation theory) describing the scattered atoms. The mean-field component evolves according to the standard time-dependent Gross–Pitaevskii (GP) equation, where the initial state is taken in the form of $\psi(\mathbf{r}, 0) = \sqrt{\rho_0(\mathbf{r})/2}\left(e^{ik_0 z} + e^{-ik_0 z}\right)$. This models an instantaneous splitting at $t = 0$ of a zero-temperature condensate in a coherent state into two halves which subsequently evolve in free space, where $\rho_0(\mathbf{r})$ is the particle number density of the initial (trapped) sample before splitting.

The fluctuating component is simulated using the stochastic counterpart of the linear operator equation [1, 5], $i\hbar\partial_t\hat{\delta}(\mathbf{r}, t) = \mathcal{H}_0(\mathbf{r}, t)\hat{\delta} + \Upsilon(\mathbf{r}, t)\hat{\delta}^\dagger$, in the positive P-representation with the vacuum initial state. Here $\mathcal{H}_0(\mathbf{r}, t) = -\frac{\hbar^2}{2m}\nabla^2 + 2U|\psi(\mathbf{r}, t)|^2 + V_{\text{BP}}(\mathbf{r}, t)$ represents the kinetic energy term, an effective mean-field potential, plus the lattice potential $V_{\text{BP}}(\mathbf{r}, t)$ imposed by the Bragg pulses, whereas $\Upsilon(\mathbf{r}, t) = U\psi(\mathbf{r}, t)^2$ is an effective coupling responsible for the spontaneous pair-production of scattered atoms. The interaction constant U is given by $U = 4\pi\hbar^2 a/m$, where m is the atomic mass, and a is the s-wave scattering length.

The Bragg pulses are realised by two interfering laser beams (assumed for simplicity to have a uniform intensity across the atomic cloud and zero relative phase) that create a periodic lattice potential $V_{\text{BP}}(\mathbf{r}, t) = \frac{1}{2}V_L(t)\cos(2\mathbf{k}_L \cdot \mathbf{r})$, where $V_L(t)$ is the lattice depth and $\mathbf{k}_L = \frac{1}{2}(\mathbf{k}_{L,2} - \mathbf{k}_{L,1})$ is the lattice vector determined by the wavevectors $\mathbf{k}_{L,i}$ ($i = 1, 2$) of the two lasers, and tuned to $|\mathbf{k}_L| = k_r$. The Bragg pulses couple momentum modes $\mathbf{k}_i$ and $\mathbf{k}_j = \mathbf{k}_i - 2\mathbf{k}_L$, satisfying momentum and energy conservation (up to a finite width due to energy-time uncertainty [6]). The lattice depth is ramped up (down) according to $V_L(t) = V_0\exp[-(t - t_2)^2/2\tau_\pi^2] + \frac{1}{2}V_0\exp[-(t - t_3)^2/2\tau_{\pi/2}^2]$, where $t_{2(3)}$ is the pulse centre, while $\tau_{\pi(\pi/2)}$ is the pulse duration which governs the

© Springer International Publishing Switzerland 2016
R.J. Lewis-Swan, *Ultracold Atoms for Foundational Tests
of Quantum Mechanics*, Springer Theses, DOI 10.1007/978-3-319-41048-7

transfer of atomic population between the targeted momentum modes: a π-pulse is defined by $\tau_\pi = \sqrt{2}\pi\hbar/V_0$ and converts the entire population from one momentum mode to the other, while a $\pi/2$-pulse is defined by $\tau_{\pi/2} = \sqrt{\pi}\hbar/(\sqrt{2}V_0)$ and converts only half of the population.

In practice, the atom–atom correlations quantifying the HOM interference are measured in position space after the low-density scattering halo expands ballistically in free space and falls under gravity onto an atom detector. The detector records the arrival times and positions of individual atoms, which is literally the case for metastable helium atoms considered here [1, 3, 4, 7, 8]. The arrival times and positions are used to reconstruct the three-dimensional velocity (momentum) distribution before expansion, as well as the atom–atom coincidences for any desired pair of momentum vectors. In our simulations and the proposed geometry of the experiment, the entire system (including the Bragg pulses) maintains reflectional symmetry about the yz-plane, with z being the vertical direction. Therefore the effect of gravity can be completely ignored as it does not introduce any asymmetry to the momentum distribution of the atoms and their correlations on the equatorial plane of the halo or indeed any other plane parallel to it.

C.2 Model Hamiltonian in Undepleted Pump Approximation

The simplest model of the collision process is for an initial homogenous condensate of fixed density ρ_0 which is treated in the undepleted pump approximation. We have previously considered this situation in Appendix A and we direct the reader therein for further details.

The dynamics of the scattered atoms in the collision halo are given by the solutions

$$\hat{a}_{\mathbf{k}}(t) = \alpha_{\mathbf{k}}(t)\hat{a}_{\mathbf{k}}(0) + \beta_{\mathbf{k}}(t)\hat{a}^\dagger_{-\mathbf{k}}(0), \tag{C.1}$$

$$\hat{a}^\dagger_{-\mathbf{k}}(t) = \beta^*_{\mathbf{k}}(t)\hat{a}_{\mathbf{k}}(0) + \alpha^*_{\mathbf{k}}(t)\hat{a}^\dagger_{-\mathbf{k}}(0), \tag{C.2}$$

where $\hat{a}_{\mathbf{k}}(t)$ $(\hat{a}^\dagger_{\mathbf{k}}(t))$ are the annihilation (creation) operators corresponding to the mode $\mathbf{k}$ in the collision halo and the coefficients are given by

$$\alpha_{\mathbf{k}}(t) = \left[\cosh\left(\sqrt{g^2 - \Delta_k^2}\, t\right) - \frac{i\Delta_k}{\sqrt{g^2 - \Delta_k^2}}\sinh\left(\sqrt{g^2 - \Delta_k^2}\, t\right)\right] e^{i\frac{\hbar|\mathbf{k}_0|^2}{2m}t}, \tag{C.3}$$

$$\beta_{\mathbf{k}}(t) = \frac{-ig}{\sqrt{g^2 - \Delta_k^2}}\, \sinh\left(\sqrt{g^2 - \Delta_k^2}\, t\right) e^{i\frac{\hbar|\mathbf{k}_0|^2}{2m}t}, \tag{C.4}$$

for an effective coupling strength $g \equiv U\rho_0/\hbar$ and $\Delta_k \equiv \hbar|\mathbf{k}|^2/2m - \hbar|\mathbf{k}_0|^2/2m$ where $\pm\mathbf{k}_0$ is the momenta of the two counter-propagating halves of the initial BEC.

These solutions are physically valid in the short-time limit, corresponding in general to less than 10 % depletion of the source condensate.

In this model, atom–atom correlations in the scattering halo can be completely characterised by

$$n_{\mathbf{k}}(t) = \langle \hat{a}_{\mathbf{k}}^{\dagger}(t)\hat{a}_{\mathbf{k}}(t)\rangle = |\beta_{\mathbf{k}}(t)|^2, \tag{C.5}$$

$$m_{\mathbf{k},-\mathbf{k}}(t) = \langle \hat{a}_{\mathbf{k}}(t)\hat{a}_{-\mathbf{k}}(t)\rangle = \alpha_{\mathbf{k}}(t)\beta_{\mathbf{k}}(t), \tag{C.6}$$

which are known as the normal and anomalous densities respectively.

Even though the undepleted pump approximation outlined in Appendix A and the Bogoliubov approach used in the numerical simulations in the main text share the same property that they both assume a constant total number of atoms in the colliding source condensates, there is an important difference between the two approaches. While the simple analytic solutions obtained above assume that the condensate *densities* remain constant as well, the Bogoliubov approach does not impose this condition. Instead, it treats the expansion of the colliding condensates in free space as prescribed by the Gross–Pitaevskii Eq. for the mean-field component. This means that the Bogoliubov counterpart of the effective coupling $g = U\rho_0/\hbar$ (see Appendix A for the source of this term) whilst being spatially dependent also becomes smaller with time as the condensate densities decrease during the expansion. Because of this difference, the analytic results to be derived and discussed in this Supplementary section can only serve for qualitative insights into the physics behind the HOM effect for matter waves, but they will not necessarily be in quantitative agreement with the numerical results presented in the main text.

C.3 Width of the HOM Dip

To estimate the width of the HOM dip after the application of Bragg pulses, we approximate the pulses as perfect mirrors and symmetric (50:50) beam-splitters over the relevant regions of the scattering halo, allowing us to model them as a series of simple linear transformations on the creation (annihilation) operators. Using Wick's theorem, we can then express the discrete-operator counterpart of the second-order correlation function considered in the main text, $g_{\mathrm{RL}}^{(2)}(t) = \langle : \hat{n}_{\mathrm{R}}(t)\hat{n}_{\mathrm{L}}(t): \rangle / \langle \hat{n}_{\mathrm{R}}(t)\rangle\langle \hat{n}_{\mathrm{L}}(t)\rangle$ at time t_4, purely in terms of the normal and anomalous densities at the end of the collision at time t_1,

$$g_{\mathrm{RL}}^{(2)}(t_4) = \frac{1}{2} + \frac{n_{\mathbf{k}_3}(t_1)^2 + n_{\mathbf{k}_5}(t_1)^2}{\left(n_{\mathbf{k}_3}(t_1) + n_{\mathbf{k}_5}(t_1)\right)^2} + \frac{|m_{\mathbf{k}_3,\mathbf{k}_4}(t_1)|^2 + |m_{\mathbf{k}_5,\mathbf{k}_6}(t_1)|^2}{2\left(n_{\mathbf{k}_3}(t_1) + n_{\mathbf{k}_5}(t_1)\right)^2}$$
$$- \frac{m_{\mathbf{k}_3,\mathbf{k}_4}^*(t_1)m_{\mathbf{k}_5,\mathbf{k}_6}(t_1)e^{i\phi} + \mathrm{h.c.}}{2\left(n_{\mathbf{k}_3}(t_1) + n_{\mathbf{k}_5}(t_1)\right)^2}, \tag{C.7}$$

where $\phi = \phi(\theta) \equiv 8\hbar|\mathbf{k}_0|^2 \Delta t_{\mathrm{free}} \sin^2(\theta/2)/m$ and Δt_{free} is the duration of free-propagation after the π-pulse.

For the HOM dip minimum at $\theta = 0$, this simple model predicts $g_{RL}^{(2)}(t_4) = 1$, whilst for sufficiently large θ, such that the momenta $\mathbf{k}_{5,6}$ lie outside the scattering halo, we find $g_{RL}^{(2)}(t_4) = 2 + 1/2n_{\mathbf{k}_3}(t_1)$. For intermediate values of θ, the full HOM dip profile is described by Eq. (C.7) and is shown in Fig. C.1 by the solid (red) curve. From the structure of Eq. (C.7), it is clear that the characteristic width of the dip will depend strongly on the widths of the densities $|m_{\mathbf{k}_5,\mathbf{k}_6}(t_1)|$ and $n_{\mathbf{k}_5}(t_1)$.

For a simple analytic estimate of the dip width we further approximate the radial profile of the halo density, Eq. (C.5), as well as of the anomalous moment, Eq. (C.6), which are both spherically symmetric, by Gaussian functions of the form $\propto \exp\left[-(k-k_0)^2/2\delta k_r^2\right]$, where $k \equiv |\mathbf{k}|$ and δk_r is the rms width. The relevant densities are then given by

$$n_{\mathbf{k}_3}(t_1) = n_{\mathbf{k}_4}(t_1) = n_0, \tag{C.8}$$

$$n_{\mathbf{k}_5}(t_1) = n_{\mathbf{k}_6}(t_1) = n_0 e^{-|\mathbf{k}_0|^2\left[1-\sqrt{5-4\cos(\theta)}\right]^2/2(\delta k_r)^2}, \tag{C.9}$$

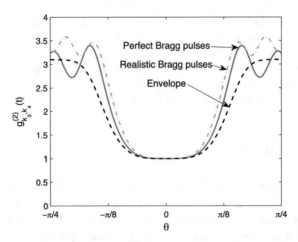

Fig. C.1 Normalised correlation function $g_{RL}^{(2)}(t_4)$ between atomic populations after the $\pi/2$-pulse. The HOM dip is realised in the simplest model [Eq. (C.7)], corresponding to a uniform BEC in the undepleted pump approximation and perfect mirrors and a symmetric beam-splitter (*full red line*). We also consider the case of off-resonant Bragg pulses (*green dot-dashed line*), corresponding to the case of an asymmetric beam-splitter (see Sect. C.5). For comparison we plot the envelope fit of Eq. (C.12) (*dashed black curve*), which shows reasonable agreement with the overall shape of the dip of Eq. (C.7). For all analytic calculations, the uniform density ρ_0 is chosen to match the peak density of the source BEC used the numerical results of the main text. The collision duration (and a matching free-propagation time), on the other hand, is chosen to be somewhat shorter ($t_1 = 30\,\mu s$) in order to result in a radial rms width of the scattering halo ($\delta k_r \simeq 0.1|\mathbf{k}_0|$) that agrees with the one obtained in pure numerical simulations. While overestimating the peak mode occupancy in the scattering halo ($n_{\mathbf{k}_3}(t_1) = 0.45$), this choice of parameters optimises the overall shape of the HOM dip as a function of the widths of the normal and anomalous densities, $n_{\mathbf{k}_6}(t_1)$ and $|m_{\mathbf{k}_5,\mathbf{k}_6}(t_1)|$ (color figure online)

$$|m_{\mathbf{k}_3,\mathbf{k}_4}(t_1)| = m_0, \tag{C.10}$$

$$|m_{\mathbf{k}_5,\mathbf{k}_6}(t_1)| = m_0 e^{-|\mathbf{k}_0|^2\left[1-\sqrt{5-4\cos(\theta)}\right]^2/2(\delta k_r)^2}. \tag{C.11}$$

Here, n_0 is the peak occupancy predicted by Eq. (C.5) and $m_0 = \sqrt{n_0(1+n_0)}$, where we have have used the identity $|m_{\mathbf{k},-\mathbf{k}}(t)|^2 = n_{\mathbf{k}}(t)[1+n_{\mathbf{k}}(t)]$. To simplify our approximation of Eq. (C.7), we impose the condition $\mathrm{Arg}\left[m^*_{\mathbf{k}_3,\mathbf{k}_4}(t_1)m_{\mathbf{k}_5,\mathbf{k}_6}(t_1)e^{i\phi}\right] = 0$, which amounts to ignoring any phase variations of the anomalous density across the scattering halo. Combined with Eqs. (C.9) and (C.11), this simplification allows us to write Eq. (C.7) in the form

$$g^{(2)}_{\mathrm{RL}}(t_4) = \frac{1}{2} + \frac{1+n_0}{2n_0}\tanh^2\left(\beta(\theta)\right) + \frac{1}{\mathrm{sech}\left(\beta(\theta)\right)+1}, \tag{C.12}$$

where

$$\beta(\theta) \equiv \frac{|\mathbf{k}_0|^2\left[1-\sqrt{5-4\cos(\theta)}\right]^2}{2(\delta k_r)^2}. \tag{C.13}$$

The second-order correlation $g^{(2)}_{\mathrm{RL}}(t_4)$, Eq. (C.12), as a function of the angle θ, has a full width at half maximum (FWHM) with respect to unity of

$$w_{dip} = 2\arccos\left[\frac{5}{4} - \frac{1}{4}\left(1 + \sqrt{\frac{2(\delta k_r)^2\beta_0}{|\mathbf{k}_0|^2}}\right)^2\right], \tag{C.14}$$

in units of radians, where

$$\beta_0 \equiv \log\left(3 + \frac{2\sqrt{1+2\left(1+\frac{1}{2n_0}\right)^2}}{1+\frac{1}{2n_0}}\right). \tag{C.15}$$

In Fig. C.1 we plot the envelope fit to the HOM dip profile, Eq. (C.12), as a dashed curve, which shows reasonable agreement with the full analytic result of Eq. (C.7) in terms of the overall shape of the dip. The discrepancies in the width of the dip are completely attributable to our assumption that $|m_{\mathbf{k}_5,\mathbf{k}_6}(t_1)|$ shares the same rms width as $n_{\mathbf{k}_5}(t_1)$, and our neglection of the phase profile $\mathrm{Arg}\left[m^*_{\mathbf{k}_3,\mathbf{k}_4}(t_1)m_{\mathbf{k}_5,\mathbf{k}_6}(t_1)e^{i\phi}\right]$. The oscillations in the wings of Eq. (C.7) are due to a combination of this phase profile and oscillations in $n_{\mathbf{k}}(t_1)$ at the spontaneous noise level for $g^2 - \Delta_k^2 < 0$ outside the scattering halo.

Lastly, by comparison of Eq. (C.7) to the phase-insensitive envelope fit in Fig. C.1, it is clear that the oscillations in the wings of Eq. (C.7) are centred on a mean-value $\overline{g^{(2)}_{\mathrm{RL}}(t_4)} = 2 + 1/2n_{\mathbf{k}_3}(t_1)$. This observation justifies our definition of dip visibility employed in the main text.

C.4 Relation Between HOM Effect and Cauchy–Schwarz Inequality

The quantum nature of the Hong–Ou–Mandel effect is commonly characterised by the visibility of the HOM dip. In this section we outline the relation between this visibility and the violation of the Cauchy–Schwarz inequality, which has been demonstrated in condensate collisions in Refs. [3, 4].

The visibility of the HOM dip is defined as $V = 1 - \min[g_{\text{RL}}^{(2)}(t_4)]/\max[g_{\text{RL}}^{(2)}(t_4)]$, where $\min[g_{\text{RL}}^{(2)}(t_4)]$ occurs for $\theta = 0$ and $\max[g_{\text{RL}}^{(2)}(t_4)]$ corresponds to sufficiently large θ such that momenta $\mathbf{k}_{5,6}$ lie outside the scattering halo. To highlight the link to the Cauchy–Schwarz inequality we evaluate these quantities by rewriting (C.7) in terms of the second-order correlations $g_{\mathbf{k},\mathbf{k}'}^{(2)}(t) = \langle \hat{a}_{\mathbf{k}}^{\dagger}(t)\hat{a}_{\mathbf{k}'}^{\dagger}(t)\hat{a}_{\mathbf{k}'}(t)\hat{a}_{\mathbf{k}}(t)\rangle/\langle \hat{a}_{\mathbf{k}}^{\dagger}(t)\hat{a}_{\mathbf{k}}(t)\rangle\langle \hat{a}_{\mathbf{k}'}^{\dagger}(t)\hat{a}_{\mathbf{k}'}(t)\rangle$ at the end of the collision at time t_1,

$$\min[g_{\text{RL}}^{(2)}(t_4)] = \frac{1}{2}g_{\mathbf{k}_3\mathbf{k}_3}^{(2)}(t_1) \tag{C.16}$$

$$\max[g_{\text{RL}}^{(2)}(t_4)] = \frac{1}{2}\left[g_{\mathbf{k}_3\mathbf{k}_4}^{(2)}(t_1) + g_{\mathbf{k}_3\mathbf{k}_3}^{(2)}(t_1)\right] \tag{C.17}$$

where we use the symmetry $g_{\mathbf{k}_3\mathbf{k}_3}^{(2)}(t_1) = g_{\mathbf{k}_4\mathbf{k}_4}^{(2)}(t_1)$.

The Cauchy–Schwarz inequality, in the context of the correlations after the collision, is given as $g_{\mathbf{k}_i,\mathbf{k}_j}^{(2)}(t_1) \leq \sqrt{g_{\mathbf{k}_i,\mathbf{k}_i}^{(2)}(t_1)g_{\mathbf{k}_j,\mathbf{k}_j}^{(2)}(t_1)}$ where we assume $n_{\mathbf{k}_i}(t_1) = n_{\mathbf{k}_j}(t_1)$. We characterise a violation of the inequality by the quantity $C = g_{\mathbf{k}_i,\mathbf{k}_j}^{(2)}(t_1)/\sqrt{g_{\mathbf{k}_i,\mathbf{k}_i}^{(2)}(t_1)g_{\mathbf{k}_j,\mathbf{k}_j}^{(2)}(t_1)} > 1$. Using this and Eqs. (C.16) and (C.17) we may quantify the visibility of the HOM dip as,

$$V = 1 - \frac{1}{1 + C}. \tag{C.18}$$

A measurement of $V > 0.5$ corresponds strictly to $C > 1$ and thus a violation of the inequality, implying the correlations between scattered atom pairs cannot be described by classical stochastic random variables [9].

C.5 Effects of Realistic Bragg Pulses

In the qualitative description of our model and the simplified undepleted pump description, we assume perfect π and $\pi/2$-pulses for all momentum components which are coupled (i.e., 100 % and 50 % transfer of atomic populations respectively). However, such perfect transfer only occurs for the momentum components, $\mathbf{k}_1$ and $\mathbf{k}_2$ (corresponding to $\theta = 0$), specifically targeted by the Bragg pulse, which satisfy the Bragg resonance condition for momentum and energy conservation. For $|\theta| > 0$, on the other hand, the coupled components $\mathbf{k}_{3(4)}$ and $\mathbf{k}_{6(5)}$ do not conserve energy and are detuned from this resonance condition, leading to a population transfer varying from the canonical definition of π and $\pi/2$-pulses. In this section we investigate the quantitative effects such off-resonant coupling has on the nature of the HOM dip.

For a simple insight we model the case of square Bragg pulses where the lattice depth is ramped on/off instantaneously, $V_L(t) = V_0 \Theta(t - t_{\mathrm{on}})[1 - \Theta(t - t_{\mathrm{off}})]$ where Θ is the Heaviside step function, and restrain coupling to momentum components separated by a single momentum kick, $\mathbf{k}_{i,j}$ where $\mathbf{k}_j = \mathbf{k}_i - 2\mathbf{k}_L$. A π-pulse is defined by the duration $\tau_\pi = 2\pi\hbar/V_0$ and a $\pi/2$-pulse by $\tau_{\pi/2} = \pi\hbar/V_0$. This model can be solved analytically (see, e.g., Ref. [10]) to give the transmission and reflection amplitudes of the pulses, and is a reasonably valid approximation to the Gaussian Bragg pulses used in numerical simulations. To compare directly we note that square and Gaussian Bragg pulses of the same lattice depth and of duration τ and τ' respectively are related by the equivalence relation $\tau = \sqrt{2\pi}\tau'$.

The collision process is again treated according to the undepleted pump model outlined in Sect. C.2; in Fig. C.1 we plot the resulting $g_{\mathrm{RL}}^{(2)}(t_4)$ for the case of realistic Bragg pulses as a dash-dotted (green) curve. We choose $\tau_\pi = \tau_{\pi/2} = \sqrt{2\pi}\tau'$ where $\tau' = 2.5\ \mu s$ matches the pulse duration used in the simulations of the main text. For small θ we find little deviation from calculations based on perfect mirror/beam-splitter transformations [shown as the solid (red) curve]; the overall shape of the dip is preserved, although there is a slight decrease in the FWHM. For large θ the effects of the off-resonant coupling become larger, resulting in a decrease in period and amplitude of the oscillations in the wings of $g_{\mathrm{RL}}^{(2)}(t_4)$. In addition, we observe the mean value in the wings, $\overline{g_{\mathrm{RL}}^{(2)}(t_4)}$, increases slightly, relative to the case of perfect mirror/beam-splitter transformations. However, the increase is sufficiently small so as not to affect our claim of a nonclassical visibility $V > 0.5$.

C.6 Impact of Realistic Bragg Pulses on Distinguishability

Beyond the quantitative changes to the structure of the HOM dip, another issue arising from off-resonant coupling relates to the treatment of path distinguishability in the scheme. In the archetypal optical HOM effect, perfect suppression of correlations between opposing output ports of the interferometer occurs only for symmetric (50:50) beam-splitters. In practice, asymmetry in the beam-splitter reflection/transmission (R/T) amplitudes provides which-way information (path distinguishability), leading to a decrease in dip visibility [11]. In this section we investigate how off-resonant coupling plays a similar role in our proposed scheme and seek to quantify the impact it may have on the visibility of the HOM dip.

In the qualitative analysis of our model, we observe that the inhomogeneous density of the scattering halo is pivotal to revealing the structure of the HOM dip. When atomic populations in components $\mathbf{k}_3$ and $\mathbf{k}_4$ are coupled to vacuum outside the populated region of the scattering halo (components $\mathbf{k}_6$ and $\mathbf{k}_5$ respectively), the paths through the beam-splitter are completely distinguishable. Such coupling corresponds to large θ, where we have demonstrated the detuning from Bragg resonance has appreciable effects. In principle, if the detuning from perfect Bragg resonance is sufficiently large such that our $\pi/2$-pulse corresponds to $|R|^2 = 1$ and $|T|^2 = 0$ (or vice versa) for the off-resonant components, the visibility of the HOM dip would be completely attributable to which-way information gained from the off-resonant coupling rather than the inhomogeneous profile of the scattering halo.

To quantify the distinguishability provided by off-resonant coupling separately to that produced by the non-uniform scattering halo, we consider an artificial model describing the scattered atoms, wherein we remove all spatial structure from Eqs. (C.5) and (C.6). The normal and anomalous densities are then completely characterised by

$$n_{\mathbf{k}}(t) = n_0, \tag{C.19}$$

$$m_{\mathbf{k},-\mathbf{k}}(t) = -i\sqrt{n_0(1 + n_0)}, \tag{C.20}$$

where n_0 is the average occupation of the modes, chosen to match the peak of Eq. (C.5). We preserve the relation $|m_{\mathbf{k},-\mathbf{k}}(t)|^2 = n_{\mathbf{k}}(t)[1 + n_{\mathbf{k}}(t)]$ and for definiteness we have chosen the phase of the anomalous density to match that of Eq. (C.6) for $\Delta_k = 0$. To be consistent with this choice of phase profile, we also neglect any free-propagation effects in this calculation.

In Fig. C.2 we plot the maximum visibility of the HOM dip for the case of a uniform halo and taking into account the off-resonant coupling of both the π and $\pi/2$-pulses (following the analytic treatment of Sect. C.5), compared to that expected for an inhomogeneous halo with perfect mirror/beam-splitter transformations (dashed line). As the visibility measure is sensitive to the mode occupation, we choose $n_0 = 0.14$ to match the numerical results of the main text. We find that shorter pulse durations limit the effects of off-resonant coupling, which is in agreement with the results of Ref. [10] where the efficiency of transfer over a broad momentum width is found to decrease with pulse duration. It is also clear in Fig. C.2 that the maximum visibility due to off-resonant coupling cannot match that expected from an inhomogeneous halo for any τ investigated. The remaining small difference between the two curves at large τ is due to our modelling of imperfect mirrors, in addition to imperfect beam-splitter.

For the Gaussian pulse scheme used in the main text ($\tau' = 2.5\ \mu$s) we calculate worst-case reflection amplitudes of $|R_\pi|^2 = 0.84$ and $|R_{\pi/2}|^2 = 0.43$ for the π and $\pi/2$-pulses respectively, corresponding to $|\theta| = \pi/4$. The which-way information gained from this is predicted to give a visibility of 70%, compared to 82% for an inhomogeneous halo. This may seem large, however, it is important to note that this only corresponds to a maximal correlation of $g_{\mathrm{RL}}^{(2)}(t_4) = 0.62(2 + 1/2n_0)$ (see Fig. C.2), whereas for an inhomogeneous scattering halo we expect an average value of $g_{\mathrm{RL}}^{(2)}(t_4) = 2 + 1/2n_0$ for large $|\theta|$. We thus conclude that, for the parameter regime simulated, full distinguishability of the paths through the interferometer and hence

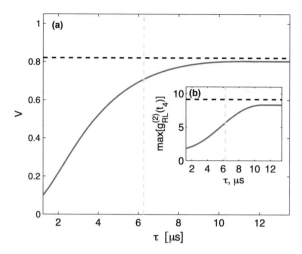

Fig. C.2 **a** Visibility of the HOM dip as a function of pulse duration $\tau_\pi = \tau_{\pi/2} = \tau$. We compare the case of a uniform scattering halo taking into account off-resonant coupling (*solid blue line*) with that calculated from the full undepleted pump model (inhomogeneous scattering halo) and perfect Bragg pulses (*dashed black line*). We indicate on the figure (*vertical dashed line*) the equivalent pulse duration for the Gaussian pulses used in the main numerical results ($\tau' = 2.5\ \mu$s), which leads to a maximum visibility of 70% ($V = 0.70$) for the case of realistic Bragg pulses, compared to 82% for the perfect case. **b** The maximum obtained $g_{\text{RL}}^{(2)}(t_4)$, corresponding to $|\theta| = \pi/4$, for the case of realistic (*solid blue line*) and perfect (*dashed black line*) Bragg pulses. The perfect case corresponds to the average value $\overline{g_{\text{RL}}^{(2)}}(t_4) = 2 + 1/2n_0$. Equivalent Gaussian pulse duration is again indicated (*vertical dashed line*) for reference (color figure online)

the magnitude of the dip visibility cannot be purely explained as a consequence of off-resonant coupling, but requires the inhomogeneity of the scattering halo to be taken into account.

References

1. Krachmalnicoff, V., et al.: Spontaneous four-wave mixing of de Broglie waves: beyond optics. Phys. Rev. Lett. **104**, 150402 (2010)
2. Deuar, P., Chwedeńczuk, J., Trippenbach, M., Ziń, P.: Bogoliubov dynamics of condensate collisions using the positive-P representation. Phys. Rev. A **83**, 063625 (2011)
3. Jaskula, J.-C., et al.: Sub-Poissonian number differences in four-wave mixing of matter waves. Phys. Rev. Lett. **105**, 190402 (2010)
4. Kheruntsyan, K.V., et al.: Violation of the Cauchy-Schwarz inequality with matter waves. Phys. Rev. Lett. **108**, 260401 (2012)
5. Ogren, M., Kheruntsyan, K.V.: Atom–atom correlations in colliding Bose-Einstein condensates. Phys. Rev. A. **79**, 021606 (2009)
6. Batelaan, H.: The Kapitza-Dirac effect. Contemp. Phys. **41**, 369–381 (2000)
7. Perrin, A., et al.: Observation of atom pairs in spontaneous four-wave mixing of two colliding Bose-Einstein condensates. Phys. Rev. Lett. **99**, 150405 (2007)

8. Vassen, W., et al.: Cold and trapped metastable noble gases. Rev. Mod. Phys. **84**, 175–210 (2012)
9. Su, C., Wódkiewicz, K.: Quantum versus stochastic or hidden-variable fluctuations in two-photon interference effects. Phys. Rev. A **44**, 6097–6108 (1991)
10. Szigeti, S.S., Debs, J.E., Hope, J.J., Robins, N.P., Close, J. D.: Why momentum width matters for atom interferometry with Bragg pulses. New J. Phys. **14**, 023009 (2012)
11. Hong, C.K., Ou, Z.Y., Mandel, L.: Measurement of subpicosecond time intervals between two photons by interference. Phys. Rev. Lett. **59**, 2044–2046 (1987)

Appendix D
Supplementary Material for Chapter 3

D.1 The Undepleted Pump Approximation and Relation to the Model of Spontaneous Parametric Down-Conversion

The simplest analytic treatment of the scheme can be made by treating the initially split-condensate in the undepleted pump approximation [1], corresponding to short collision durations such that the number of scattered atoms is only a small fraction of the source condensate (generally less than 10 %). An examination of this model is given in Appendix A and we direct the reader therein for specific details. We treat the laser-induced π and $\pi/2$ Bragg pulses, which are characterised by a momentum kick $2\mathbf{k}_L \equiv \mathbf{k}_{L,2} - \mathbf{k}_{L,1} = \mathbf{k}_3 - \mathbf{k}_1 = \mathbf{k}_2 - \mathbf{k}_4$ (see Appendix B for further details), as perfect mirrors and beam splitters (i.e., simple linear transformations) applied at t_2 and t_4 respectively (see main text for definitions) and then invoking Wick's theorem, the second-order correlation function between the relevant pairs of detectors (chosen for definiteness to be equal to $t_4 = t_3 + 4\tau_{\pi/2}$ in our simulations) can be written as

$$G^{(2)}(\mathbf{k}_1, \mathbf{k}_2, t_4) = G^{(2)}(\mathbf{k}_3, \mathbf{k}_4, t_4) = n(\mathbf{k}_1, t_1)^2 + \frac{|m(\mathbf{k}_1, \mathbf{k}_2, t_1)|^2}{2} [1 - \cos(\phi_R - \phi_L)],$$

(D.1)

$$G^{(2)}(\mathbf{k}_1, \mathbf{k}_4, t_4) = G^{(2)}(\mathbf{k}_2, \mathbf{k}_3, t_4) = n(\mathbf{k}_1, t_1)^2 + \frac{|m(\mathbf{k}_1, \mathbf{k}_2, t_1)|^2}{2} [1 + \cos(\phi_R - \phi_L)],$$

(D.2)

where $n(\mathbf{k}, t_1) = \langle \hat{a}^\dagger(\mathbf{k}, t_1)\hat{a}(\mathbf{k}, t_1) \rangle$ is the average momentum-space density of scattered atoms after the collision at time t_1, which is equal for the targeted modes $\mathbf{k}_1, \mathbf{k}_2, \mathbf{k}_3$ and $\mathbf{k}_4$, and $m(\mathbf{k}, \mathbf{k}', t_1) = \langle \hat{a}(\mathbf{k}, t_1)\hat{a}(\mathbf{k}', t_1) \rangle$ is the average anomalous moment. Choosing $\phi_L = 0$, $\phi'_L = \pi/2$, $\phi_R = \pi/4$ and $\phi'_R = 3\pi/4$ to maximise the CHSH-Bell parameter S (defined as per the main text) we find the result

© Springer International Publishing Switzerland 2016
R.J. Lewis-Swan, *Ultracold Atoms for Foundational Tests of Quantum Mechanics*, Springer Theses, DOI 10.1007/978-3-319-41048-7

$$S = 2\sqrt{2}\frac{|m(\mathbf{k}_1, \mathbf{k}_2, t_1)|^2}{2n(\mathbf{k}_1, t_1)^2 + |m(\mathbf{k}_1, \mathbf{k}_2, t_1)|^2}. \tag{D.3}$$

For a maximal violation, with $S = 2\sqrt{2}$, one requires the anomalous moment to satisfy $|m(\mathbf{k}_1, \mathbf{k}_2, t_1)|^2 \gg n(\mathbf{k}_1, t_1)^2$, corresponding to strong correlations between atoms scattered to diametrically opposite momentum modes.

The anomalous moment is maximised for the case of a homogeneous BEC in a finite box [1, 2], where the discrete-mode counterpart of $m(\mathbf{k}, -\mathbf{k})$ satisfies $|m_{\mathbf{k}, -\mathbf{k}}|^2 = n_{\mathbf{k}}(1 + n_{\mathbf{k}})$ [2]—just like in the simple four-mode model of parametric down-conversion discussed in the main text, thus giving the result of Eq. (3.4), with $n = n_{\mathbf{k}_i}$ ($i = 1, 2, 3, 4$) being the average mode occupation of the scattering halo after the collision, which are all equal in this approximation.

D.2   Gaussian-Fit Analytic Model of Correlation Functions

Beyond the simple treatment of the previous section, we can develop a more sophisticated model of the CHSH-Bell parameter whilst also taking into account the finite detector resolution of experiments [3]. We calculate integrated pair-correlation functions and the ensuing CHSH-Bell parameter by using a Gaussian-fit analytic model, similar to that used previously in Ref. [4] to model a violation of the Cauchy–Schwarz inequality in condensate collisions. The underlying assumption of the model is that the second-order correlation function after the collision is well approximated by a Gaussian $G^{(2)}(\mathbf{k}, \mathbf{k}', t_1) = n^2(1 + h\prod_d \exp[-(k_d + k'_d)^2/2\sigma_d^2])$ for $\mathbf{k} \simeq -\mathbf{k}'$ and $n = n(\mathbf{k}) = n(\mathbf{k}')$ is the density of scattered atoms. The correlation is then characterised by two parameters: the height, h, above the background level and the correlation width σ_d.

To derive an expression for S we first consider the form of the integrated pair-correlation functions after the application of the $\pi/2$-pulse,

$$C_{ij} = \langle \hat{N}_i \hat{N}_j \rangle = \int_{\mathcal{V}(\mathbf{k}_i)} d^3\mathbf{k} \int_{\mathcal{V}(\mathbf{k}_j)} d^3\mathbf{k}' G^{(2)}\left(\mathbf{k}, \mathbf{k}', t_4\right), \tag{D.4}$$

where the integration bins are of dimension Δk_d ($d = x, y, z$) and volume $\mathcal{V}(\mathbf{k}_i) = \prod_d \Delta k_d$ centered around the targeted momenta $\mathbf{k}_i$ ($i = 1, 2, 3, 4$). Without loss of generality we consider the form of the correlation C_{12}, with the remaining pair-correlation functions C_{ij} being calculated in a similar manner. Treating the Bragg pulses as idealised mirrors and beam-splitters which act instantaneously, meaning we may set $t_2 = t_1$ and $t_4 = t_3$, we may write the generalised form of Eq. (D.1) as

$$
\begin{aligned}
G^{(2)}(\mathbf{k}, \mathbf{k}', t_4) = \frac{1}{4} \Big\{ & 4n(\mathbf{k}, t_2)^2 + |m(\mathbf{k}, \mathbf{k}', t_2)|^2 + |m(\mathbf{k} - 2\mathbf{k}_L, \mathbf{k}' + 2\mathbf{k}_L, t_2)|^2 \\
& - \Big[m(\mathbf{k} - 2\mathbf{k}_L, \mathbf{k}' + 2\mathbf{k}_L, t_2)^* m(\mathbf{k}, \mathbf{k}', t_2) e^{-i(\phi_L - \phi_R)} \\
& \times e^{-i\frac{\hbar}{2m}(|\mathbf{k}|^2 + |\mathbf{k}'|^2 - |\mathbf{k} - 2\mathbf{k}_L|^2 - |\mathbf{k}' + 2\mathbf{k}_L|^2)\Delta t_{\text{free}}} \Big] \\
& - \Big[m(\mathbf{k}, \mathbf{k}', t_2)^* m(\mathbf{k} - 2\mathbf{k}_L, \mathbf{k}' + 2\mathbf{k}_L, t_2) e^{i(\phi_L - \phi_R)} \\
& \times e^{i\frac{\hbar}{2m}(|\mathbf{k}|^2 + |\mathbf{k}'|^2 - |\mathbf{k} - 2\mathbf{k}_L|^2 - |\mathbf{k}' + 2\mathbf{k}_L|^2)\Delta t_{\text{free}}} \Big] \Big\}.
\end{aligned} \tag{D.5}
$$

where $\mathbf{k} \in \mathcal{V}(\mathbf{k}_1)$ and $\mathbf{k}' \in \mathcal{V}(\mathbf{k}_2)$ and $\Delta t_{\text{free}} \equiv t_3 - t_2$ is defined as the duration of free-propagation between the π and $\pi/2$ Bragg pulses. Having invoked Wick's theorem in Eq. (D.5), we may recognize that assuming the correlation function $G^{(2)}(\mathbf{k}, \mathbf{k}', t_1)$ is a Gaussian function translates to the assumption that we may model the anomalous moment as

$$
m\left(\mathbf{k}, \mathbf{k}', t_2\right) \equiv \bar{n}\sqrt{h} e^{i\theta(\mathbf{k}, \mathbf{k}', t_2)} \prod_d e^{-(k_d + k_d')^2/4\sigma_d^2}, \tag{D.6}
$$

where the density of scattered atoms is assumed to be approximately homogeneous across the integration volumes and is given by the average $\bar{n}$. The argument $\theta(\mathbf{k}, \mathbf{k}', t_2)$ of the complex anomalous moment is dependent on the specific model chosen for the collision, which we will elaborate upon momentarily.

Substituting Eq. (D.6) into Eq. (D.5) gives the more recognizable form

$$
\begin{aligned}
G^{(2)}(\mathbf{k}, \mathbf{k}', t_2) = \bar{n}^2 + \frac{\bar{n}^2 h}{2} \prod_d & \exp[-(k_d + k_d')^2/2\sigma_d^2] \\
& \times \left\{ 1 - \cos\left[\phi_L - \phi_R + \varphi(\mathbf{k}, \mathbf{k}')\right] \right\},
\end{aligned} \tag{D.7}
$$

where

$$
\begin{aligned}
\varphi(\mathbf{k}, \mathbf{k}') = \theta\left(\mathbf{k} - 2\mathbf{k}_L, \mathbf{k}' + 2\mathbf{k}_L, t_2\right) - \theta\left(\mathbf{k}, \mathbf{k}', t_2\right) \\
+ \frac{\hbar}{2m}\left(|\mathbf{k}|^2 + |\mathbf{k}'|^2 - |\mathbf{k} - 2\mathbf{k}_L|^2 - |\mathbf{k}' + 2\mathbf{k}_L|^2\right)\Delta t_{\text{free}}.
\end{aligned} \tag{D.8}
$$

In comparison to the simple toy-model of Eq. (D.1) the most important new feature of Eq. (D.7) is the addition of $\varphi(\mathbf{k}, \mathbf{k}')$, which acts as a momentum-dependent drift in the phase settings ϕ_L and ϕ_R. As the phase settings are chosen to maximise the CHSH-Bell parameter, this new term can thus lead to a reduction in S. Composed of a free-propagation component and a dependence on the argument of the anomalous moment such an effect is similar to the phase-dispersion of two-color photons in an earlier optical experiment of Rarity and Tapster [5]

To investigate the impact of this new term and to evaluate the integral in Eq. (D.4) one must know the form of $\varphi(\mathbf{k}, \mathbf{k}')$, which in turn explicitly depends on the argument $\theta(\mathbf{k}, \mathbf{k}', t_2)$ of the anomalous moment. In general, this is not trivial as it requires an analytic solution of the anomalous moment from an appropriate model for the collision. To this end, we supplement our simple Gaussian-fit model by utilizing a solution of the anomalous moment based on a perturbative approach, previously used with success in Ref. [6] (albeit for a different collision geometry—the BECs were split along the x-axis). Similar to the numerical treatment, this model takes into account the evolution of the spatial overlap of the split condensate wavepackets, however, it does not account for the spatial expansion of the condensates once released from the initial trap.

To give a tractable form of the anomalous moment we approximate the initial mean-field of the unsplit condensate as a Gaussian $\psi_0(\mathbf{x}) = \sqrt{\rho_0} \prod_d e^{-x_d^2/2\sigma_{g,d}^2}$ with peak density ρ_0 and rms widths $\sigma_{g,d}$ for $d = x, y, z$. The calculation of the anomalous moment is then straightforward and involves treating the wavefunction of the scattered atoms with a perturbative expansion to low order. For a full derivation of the model we refer the reader to Appendix A Ref. [6]. In our solution we may make the approximation that the box sizes are sufficiently small such that $|\mathbf{k} - \mathbf{k}_1| \ll |\mathbf{k}_0|$ and $|\mathbf{k}' - \mathbf{k}_2| \ll |\mathbf{k}_0|$ and assume the condensates are completely spatially separated before applying the π pulse, corresponding to $t_2/\tau_s \gg 1$ where $\tau_s = m\sigma_{g,z}/\hbar|\mathbf{k}_0|$ is the time-scale of separation. Under these limits the argument of the anomalous moment may be written as

$$\theta\left(\mathbf{k}, \mathbf{k}', t_2\right) \simeq -\frac{\hbar}{2m} \left(|\mathbf{k}|^2 + |\mathbf{k}'|^2\right) t_2 + \frac{\sigma_{g,z}}{\sqrt{\pi}|\mathbf{k}_0|} \left(\frac{|\mathbf{k}|^2 + |\mathbf{k}'|^2}{2} - |\mathbf{k}_0|^2\right), \quad \text{(D.9)}$$

which thus allows us to write the phase drift as

$$\varphi(\mathbf{k}, \mathbf{k}') = \left[8|\mathbf{k}_L|^2 - 4\mathbf{k}_L \cdot \left(\mathbf{k} - \mathbf{k}'\right)\right] \left[\frac{\hbar}{2m} \left(\Delta t_{\text{free}} - t_2\right) + \frac{\sigma_{g,z}}{2|\mathbf{k}_0|\sqrt{\pi}}\right]. \quad \text{(D.10)}$$

Using the form of Eq. (D.10) and noting that our Bragg pulses couple only along the k_y-axis it is straightforward to evaluate the integral of Eq. (D.4),

$$C_{12} = \bar{n}^2 \prod_d (\Delta k_d)^2 + \frac{\bar{n}^2 h}{2} \prod_d \sigma_d \alpha_d - \frac{\bar{n}^2 h}{2} \left(\prod_d \sigma_d\right) \alpha_x \alpha_z \beta_y \cos\left(\phi_L - \phi_R\right),$$

$$\text{(D.11)}$$

where $\alpha_d \equiv (e^{-2\lambda_d^2} - 1) + \sqrt{2\pi}\lambda_d \text{erf}(\sqrt{2}\lambda_d)$, $\lambda_d \equiv \Delta k_d/2\sigma_d$, and

$$
\beta_y \equiv i \sqrt{\frac{\pi}{2}} \frac{e^{-8A^2|\mathbf{k}_L|^2\sigma_y^2}}{4A|\mathbf{k}_L|} \left[e^{-i4A|\mathbf{k}_L|\Delta k_y} \operatorname{erf}\left(\frac{\Delta k_y + i4A|\mathbf{k}_L|\sigma_y^2}{\sqrt{2}\sigma_y} \right) \right.
$$

$$
- e^{i4A|\mathbf{k}_L|\Delta k_y} \operatorname{erf}\left(\frac{\Delta k_y - i4A|\mathbf{k}_L|\sigma_y^2}{\sqrt{2}\sigma_y} \right)
$$

$$
\left. + 2\cos\left(4A|\mathbf{k}_L|\Delta k_y\right) \operatorname{erf}\left(i2\sqrt{2}A|\mathbf{k}_L|\sigma_y \right) \right], \tag{D.12}
$$

with $A \equiv \hbar(\Delta t_{\text{free}} - t_2)/2m + \sigma_{g,z}/2k_0\sqrt{\pi}$. One can then calculate the remaining correlation functions C_{ij} in a similar fashion to find the correlation coefficient

$$
E(\phi_L, \phi_R) = \frac{C_{14} + C_{23} - C_{12} - C_{34}}{C_{14} + C_{23} + C_{12} + C_{34}} \Bigg|_{(\phi_L, \phi_R)}
$$

$$
= \frac{h\alpha_x\beta_y\alpha_z}{h\prod_d \alpha_d + 2\prod_d (\lambda_d)^2} \cos(\phi_L - \phi_R). \tag{D.13}
$$

The CHSH-Bell parameter is finally given by

$$
S = 2\sqrt{2} \left| \frac{h\alpha_x\beta_y\alpha_z}{h\prod_d \alpha_d + 2\prod_d (\lambda_d)^2} \right|. \tag{D.14}
$$

An important result of this model is the prediction that there exists an optimal free-propagation duration between the π and $\pi/2$ Bragg pulses,

$$
\Delta t_{\text{free}} = t_2 - \frac{m\sigma_{g,z}}{\hbar k_0\sqrt{\pi}}, \tag{D.15}
$$

for which $\varphi(\mathbf{k}, \mathbf{k}') = 0$ in Eq. (D.8) for all $\mathbf{k} \in \mathcal{V}(\mathbf{k}_1)$ and $\mathbf{k}' \in \mathcal{V}(\mathbf{k}_2)$ and thus the phase settings retain their original values throughout the integration bin. This corresponds to $A = 0$ in Eq. (D.12) and we then find $\beta_y = \alpha_y$. Equation (D.14) is maximised under this condition and it transforms to

$$
S = 2\sqrt{2} \frac{h\prod_d \alpha_d}{h\prod_d \alpha_d + 2\prod_d (\lambda_d)^2}, \tag{D.16}
$$

where the dependence on box-size is now characterised completely by the relative quantity $\lambda_d = \Delta k_d/2\sigma_d$ for all directions, rather than the absolute length scale Δk_y as in Eq. (D.14) along the y-axis.

In Fig. D.1a we plot Eq. (D.14) as a function of Δt_{free} and Δk_y for the case of an initial BEC of $N = 1.9 \times 10^4$ atoms to illustrate the effects of the phase drift. As inputs to the model, the correlation height h and correlation widths σ_d are extracted from the numerical data at t_1, whilst the rms width $\sigma_{g,z}$ is chosen by fitting the numerically calculated trapped condensate to a Gaussian. For Δt_{free} satisfying Eq. (D.15), S

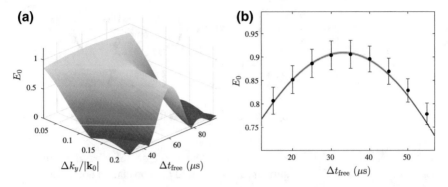

Fig. D.1 a Correlation amplitude E_0 predicted by Gaussian-fit model [Eq. (D.14)] as a function of the integration bin size Δk_y and the free propagation time Δt_{free}. Calculations were performed for an initial condensate of $N = 1.9 \times 10^4$ atoms and other parameters as per the main text with h and σ_d extracted from the stochastic numerical results. The central ridge corresponds to Eq. (D.16) where the phase drift term $\varphi(\mathbf{k}, \mathbf{k}')$ is eliminated. **b** Amplitude of the correlation function E_0 as a function of free propagation time Δt_{free} for an integration volume $(\Delta k_x, \Delta k_y, \Delta k_z) = (0.052, 0.53, 0.47)\,\mu\text{m}^{-1}$ and simulation parameters are as per (**a**). The predictions of the Gaussian-fit analytic model Eq. (D.13) (*grey shaded region*) are compared to the numerical results from stochastic simulations (*black circles*). The error bars on data points indicate the stochastic sampling error of two standard deviations obtained from ~ 800 trajectories, whilst for the analytic prediction the uncertainty in E_0 (*shaded region*) is due to the uncertainty in the values h and σ_d extracted from the numerical simulations (color figure online)

retains the maximal violation of Eq. (D.16) with the strength only declining due to a dilution of the correlation as the integration box-size Δk_y increases. However, for Δt_{free} away from the optimal value one sees that an increase in the box-size leads to a rapid decrease in S due to rapid drift of the phase-settings rather than a dilution of correlation. One can see this by noting that large Δk_y implies the term $8|\mathbf{k}_L|^2 - 4|\mathbf{k}_L| \cdot (\mathbf{k} - \mathbf{k}')$ in Eq. (D.8) will take large values near the edge of the integration volume and $\varphi(\mathbf{k}, \mathbf{k}')$ is scaled by this factor, leading to large deviations from the optimal phase-settings. This is important as it demonstrates that for poor experimental resolution even small perturbations away from the optimal Δt_{free} can lead to a quick loss of Bell violation.

Figure D.1b shows results of stochastic numerical simulations for the amplitude of the correlation function E_0, where $E(\phi_L, \phi_R) \equiv E_0 \cos(\phi_L - \phi_R)$, as a function of Δt_{free} for the same initial BEC. We compare these results to the predictions of Eq. (D.13) to investigate the applicability of the Gaussian-fit model to a realistic system. We find excellent agreement, not only for the maximum attained correlation strength but also for the predicted optimal Δt_{free}. The quantitative match to theory also implies that the underlying model for $\varphi(\mathbf{k}, \mathbf{k}')$ is a good approximation to the form in the numerical simulations, although this is expected to break down for larger integration volumes where the assumptions for $\varphi(\mathbf{k}, \mathbf{k}')$ in Eq. (D.10) are no longer satisfied.

Fig. D.2 Optimal free propagation time Δt_{free} for a range of initial BEC atom number. Numerical results (*black circles*) are compared to the prediction of Eq. (D.15) from the perturbative model (*dashed line*). The range of N in the initial BECs corresponds to those in the main text, whilst the integration volume is the same as Fig. D.1b

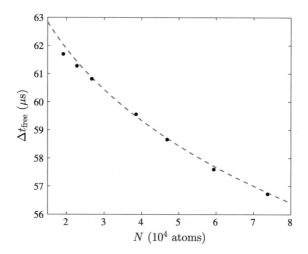

As the chosen phase angles ϕ_L and ϕ_R are shown to be unaffected in the final form of E in Eq. (D.13), it is sufficient to numerically optimise E_0 as a function of Δt_{free} to maximise the Bell violation. In Fig. D.2 we plot the optimal Δt_{free} for a variety of initial BEC atom number determined from numerical calculations and compare these to the prediction of Eq. (D.15). Once again we find good quantitative agreement between the numeric and analytic methods. The numerically determined optimal Δt_{free} here are used in the simulations of the main text to define the timing of the application of the $\pi/2$-pulse.

References

1. Perrin, A., et al.: Atomic four-wave mixing via condensate collisions. New J. Phys. **10**, 045021 (2008)
2. Savage, C.M., Schwenn, P.E., Kheruntsyan, K.V.: First-principles quantum simulations of dissociation of molecular condensates: Atom correlations in momentum space. Phys. Rev. A **74**, 033620 (2006)
3. Perrin, A., et al.: Observation of atom pairs in spontaneous four-wave mixing of two colliding Bose-Einstein condensates. Phys. Rev. Lett. **99**, 150405 (2007)
4. Kheruntsyan, K.V., et al.: Violation of the Cauchy-Schwarz inequality with matter waves. Phys. Rev. Lett. **108**, 260401 (2012)
5. Rarity, J.G., Tapster, P.R.: Two-color photons and nonlocality in fourth-order interference. Phys. Rev. A **41**, 5139–5146 (1990)
6. Chwedeńczuk, J., et al.: Pair correlations of scattered atoms from two colliding Bose-Einstein condensates: Perturbative approach. Phys. Rev. A **78**, 053605 (2008)

Appendix E
Supplementary Material for Chapter 4

E.1 Simple Results in the Undepleted Pump Approximation

To invoke the undepleted pump approximation, we assume that the pump mode is initially in a coherent state with an amplitude $\alpha_0(0) = \sqrt{N_0}$ (which we choose to be real without loss of generality) and that it does not change with time. By additionally choosing the quadratic Zeeman effect to be phase-matched ($q = gN_0$), we can reduce the model Hamiltonian to that of optical parametric-down conversion [1], $\hat{H} = \hbar\chi(\hat{a}_1^\dagger\hat{a}_{-1}^\dagger + h.c.)$, in which $\chi = gN_0$. The Heisenberg equations of motion following from this are $d\hat{a}_{\pm 1}/d\tau = -iN_0\hat{a}_{\mp 1}^\dagger$, where $\tau = gt$ is a dimensionless time. Solutions to these equations are given by

$$\hat{a}_{\pm 1}(\tau) = \cosh(N_0\tau)\hat{a}_{\pm 1}(0) - i\sinh(N_0\tau)\hat{a}_{\mp 1}^\dagger(0), \tag{E.1}$$

which are physically valid in the short-time limit, generally corresponding to less than 10 % depletion of the pump mode occupation.

Considering specific initial states for the signal and idler modes, these solutions can be used to calculate expectation values of various quantum mechanical operators and observables. For example, for a thermal initial state with an equal population in both modes, $\langle\hat{a}_1^\dagger(0)\hat{a}_1(0)\rangle = \langle\hat{a}_{-1}^\dagger(0)\hat{a}_{-1}(0)\rangle \equiv \bar{n}_{th}$, the subsequent evolution of the mode populations is given by

$$\langle\hat{a}_{\pm 1}^\dagger(\tau)\hat{a}_{\pm 1}(\tau)\rangle = \sinh^2(N_0\tau)[1 + 2\bar{n}_{th}] + \bar{n}_{th}, \tag{E.2}$$

whereas the anomalous moments evolve according to

$$\langle\hat{a}_{\pm 1}(\tau)\hat{a}_{\mp}(\tau)\rangle = -i\sinh(N_0\tau)\cosh(N_0\tau)[1 + 2\bar{n}_{th}]. \tag{E.3}$$

Similarly, the EPR entanglement parameter is found to be given by

© Springer International Publishing Switzerland 2016
R.J. Lewis-Swan, *Ultracold Atoms for Foundational Tests of Quantum Mechanics*, Springer Theses, DOI 10.1007/978-3-319-41048-7

$$\Upsilon \cong \left[\frac{(1 + 2\bar{n}_{th})^2 + \frac{1}{N_0} [(1 + 2\bar{n}_{th}) \cosh{(2N_0\tau)} - 1] [2 (1 + 2\bar{n}_{th}) \cosh{(2N_0\tau)} - 1]}{(1 + 2\bar{n}_{th}) \cosh{(2N_0\tau)} - \frac{1}{N_0} [(1 + 2\bar{n}_{th}) \cosh{(2N_0\tau)} - 1]^2} \right]^2,$$

$$(E.4)$$

where we have assumed $N_0 \gg 1$. The minimum value of this quantity (with respect to time τ) gives the maximal violation of the EPR criterion,

$$\Upsilon_{min} \cong \left[\frac{\sqrt{2N_0}}{\sqrt{\frac{1}{2}N_0} - (1 + 2\bar{n}_{th}) - \frac{1}{2N_0} \left[(1 + 2\bar{n}_{th})^3 - \left((1 + 2\bar{n}_{th})^2 + 1 \right) \sqrt{2N_0} \right]} - 2 \right]^2,$$

$$(E.5)$$

which is achieved at the optimal time

$$\tau_{min} = \frac{1}{2N_0} \text{arccosh} \left[-\frac{1}{2} (1 + 2\bar{n}_{th}) + \frac{1}{2} \sqrt{(1 + 2\bar{n}_{th})^2 + 2N_0} \right].$$

$$(E.6)$$

From Eq. (E.5) we also determine the maximum allowable thermal population before EPR entanglement is lost. By numerical analysis we find a maximum seed of $(\bar{n}_{th})_{max} \simeq 0.05 N_0^{2/3}$ in the range $100 \leqslant N_0 \leqslant 400$. We find this compares reasonably with the results of full numerical simulations, which predict $(\bar{n}_{th})_{max} \simeq 0.06 N_0^{11/20}$.

Furthermore we may also calculate the minimum two-mode quadrature variance,

$$\Delta^2 X_- = 2(1 + 2\bar{n}_{th})[\cosh(2N_0\tau) - \sinh(2N_0\tau)],$$

$$(E.7)$$

and the inter-mode inseparability parameter (see main text),

$$\Sigma \Delta_2^2 / \Sigma \Delta_1^2 = 1 - \tanh(2N_0\tau).$$

$$(E.8)$$

Despite their limited applicability and the quantitative disagreement with the numerical results, the analytic predictions of the undepleted pump approximation give useful insights into the qualitative aspects of different measures of entanglement. For example, to leading order, Eqs. (E.4) and (E.7) predict, respectively, quadratic and linear growth of the EPR entanglement parameter and two-mode squeezing with the thermal seed $\bar{n}_{th}$, whereas the inter-mode inseparability, Eq. (E.8), is insensitive to $\bar{n}_{th}$. The predictions for EPR entanglement and two-mode squeezing are in qualitative agreement with the numerical results discussed in the main text, whilst we find weak linear growth with $\bar{n}_{th}$ emerges for inter-mode inseparability due to depletion of the pump. These qualitative predictions highlight the lower tolerance and higher sensitivity of the EPR entanglement to thermal noise.

Reference

1. Walls, D.F., Milburn, G.J.: Quantum Optics, 2nd edn. Springer, Berlin (2008)

Printed in the United States
By Bookmasters